Trace Elements,
Micronutrients, and Free Radicals

Contemporary Issues
in Biomedicine, Ethics, and Society

Trace Elements, Micronutrients, and Free Radicals, edited by *Ivor E. Dreosti, 1991*

Beyond Baby M: *Ethical Issues in New Reproductive Techniques,* edited by *Dianne M. Bartels, Reinhard Priester, Dorothy E. Vawter, and Arthur L. Caplan, 1989*

Reproductive Laws for the 1990s, edited by *Sherrill Cohen* and *Nadine Taub, 1989*

The Nature of Clinical Ethics, edited by *Barry Hoffmaster, Benjamin Freedman,* and *Gwen Fraser,* 1988

What Is a Person?, edited by *Michael F. Goodman, 1988*

Advocacy in Health Care, edited by *Joan H. Marks, 1986*

Which Babies Shall Live?, edited by *Thomas H. Murray* and *Arthur L. Caplan, 1985*

Feeling Good and Doing Better, edited by *Thomas H. Murray, Willard Gaylin,* and *Ruth Macklin, 1984*

Ethics and Animals, edited by *Harlan B. Miller* and *William H. Williams, 1983*

Profits and Professions, edited by *Wade L. Robison, Michael S. Pritchard,* and *Joseph Ellin, 1983*

Visions of Women, edited by *Linda A. Bell, 1983*

Medical Genetics Casebook, by *Colleen Clements, 1982*

Who Decides?, edited by *Nora K. Bell, 1982*

The Custom-Made Child?, edited by *Helen B. Holmes, Betty B. Hoskins,* and *Michael Gross, 1980*

Medical Responsibility, edited by *Wade L. Robison* and *Michael S. Pritchard, 1979*

Contemporary Issues in Biomedical Ethics, edited by *John W. Davis, Barry Hoffmaster,* and *Sarah Shorten, 1979*

Trace Elements, Micronutrients, and Free Radicals

Edited by

Ivor E. Dreosti

Division of Human Nutrition, CSIRO, Adelaide, Australia

Humana Press • Totowa, New Jersey

Library of Congress Cataloging-in-Publication Data

Trace elements, micronutrients, and free radicals / edited by Ivor E. Dreosti.
   p. cm.
  Includes bibliographical references and index.
  ISBN 0-89603-188-8
  1. Free radicals (Chemistry)—Physiological effect. 2. Active oxygen; in the body. 3. Trace elements in human nutrition.
  I. Dreosti, Ivor E.
RB170.T73 1991
612'.01524—dc20   91-13721
                                                       CIP

Preface

It is only recently that the natural occurrence of free radicals in biological tissue has become widely accepted, and that the suspicion with which biologists previously viewed the free radicals of radiation chemistry has been placed in a broader perspective. Now, oxygen-derived free radicals are considered respectable biochemical intermediates, given always the caveat that unwanted tissue damage may arise if these active species are produced in such abundance that they overwhelm the natural antioxidant and free-radical defense mechanisms, or if these systems have become hypoeffective. Many factors, including several dietary manipulations, can lead to elevated production of superoxide and may result in free radical overload, whereas a deficiency of those micronutrients associated with the antioxidant defense mechanisms may result in substantially diminished antioxidant capacity.

By now, antioxidants have become a household word and almost everyone is aware of their importance in protecting the body against attack by active oxygen species. Indeed, it is a paradox of nature that oxygen, which is so essential to sustain aerobic life, ultimately contributes to its destruction. Not surprisingly, recognition of this dilemma has generated a spate of antioxidant strategies intended to reduce the risk of tissue damage by rampant oxygen radicals, some sadly based less on science than on speculation. Nevertheless, it is clear that free radical damage occurs continuously in aerobic organisms and that complex defense systems exist to minimize it. Also, there can be no question concerning the signal contribution made by several micronutrients to these defenses, nor about the participation of free radicals in the pathogenesis of several degenerative diseases in humans—although much remains unanswered. At a nutritional level, it is now well established that dietary intakes of the antioxidant micronutrients at levels below the requirement will raise the risk of free radical damage, although the value of massive supplementation of these dietary factors in the pursuit of supranormal health has not been proven.

The present volume addresses the issue of free radical damage and micronutrient protection in humans and animals, and seeks to bridge the gap between the separate disciplines of free radical chemistry, cellular biology, and micronutrient nutrition in order to provide the broader basis from which the suggested roles of pro- and antioxidants in human disease and aging will need to be studied. Emphasis has been placed primarily on fundamental scientific aspects of the question rather than on the possible use of micronutrients as therapeutic agents for the treatment of human disorders. Nevertheless, recognizing that this objective may well represent a substantial goal in antioxidant research, several later chapters have included some discussion of these matters.

It is our hope that this volume will provide a comprehensive and current discussion of the role of micronutrients in antioxidant protection, and we trust that it will, for some time at least, provide a *vade mecum* for those scientists working in, or entering this exciting new research area. We acknowledge with pleasure and thanks the participation of our distinguished panel of contributors and the support of The Humana Press.

Ivor E. Dreosti

Contents

v   Preface
         Ivor E. Dreosti

1   **1** • Free Radicals in Biological Systems
         Martyn C. R. Symons

25   **2** • Free Radicals as Mediators of Tissue Injury
         Rolando Del Maestro

53   **3** • Free Radical Biology of Iron
         Christine C. Winterbourn

77   **4** • Dietary Prooxidants
         Mario Umberto Dianzani

107   **5** • Essential Trace Elements in Antioxidant Processes
         Sheri Zidenberg-Cherr and Carl L. Keen

129   **6** • Vitamins and Related Dietary Antioxidants
         Ching K. Chow

149   **7** • Free Radical Pathology and the Genome
         Ivor E. Dreosti

171   **8** • The Role of Free Radicals in Cancer and Aging
         T. Mark Florence

199   **9** • Free Radicals and Malnutrition
         *Michael H. N. Golden, Dan D. Ramdath,
         and Barbara E. Golden*

223   Conclusions
         Ivor E. Dreosti

227   Index

Contributors

Ching K. Chow • Department of Nutrition and Food Science, University of Kentucky, Lexington, KY

Rolando Del Maestro • Brain Research Laboratories, University of Western Ontario, Ontario, Canada

Mario Umberto Dianzani • Institute of General Pathology, University of Turin, Turin, Italy

Ivor E. Dreosti • Division of Human Nutrition, CSIRO, Adelaide, Australia

T. Mark Florence • Division of Energy Resources, CSIRO, Adelaide, Australia (currently with the Centre for Environmental and Health Sciences)

Barbara E. Golden • Tropical Metabolism Research Unit, University of the West Indies, Kingston, Jamaica

Michael H. N. Golden • Tropical Metabolism Research Unit, University of the West Indies, Kingston, Jamaica

Carl L. Keen • Department of Nutrition, University of California, Davis, CA

Dan D. Ramdath • University of the West Indies, Kingston, Jamaica

Martyn C. R. Symons • Department of Chemistry, The University, Leicester, United Kingdom

Christine C. Winterbourn • Department of Pathology, Christchurch Hospital, Christchurch, New Zealand

Sheri Zidenberg-Cherr • Department of Nutrition, University of California, Davis, CA

CHAPTER 1

Free Radicals
in Biological Systems

MARTYN C. R. SYMONS

ABSTRACT

In this introductory chapter, the nature of 'free radicals' is discussed, and methods for detection and identification are outlined,
with particular reference to electron spin resonance (ESR) spectroscopy. Common methods for generating radicals are summarized.

A range of simple inorganic radicals of particular biological significance are discussed, including the oxygen radicals, O_2-, HO_2.
and OH·. The use of spin-traps in ESR studies of these species is
outlined. After discussing a few important organic radicals, the chapter finishes with a brief review of nitroxide radicals and their uses.

1. BACKGROUND

This is an introductory chapter for those not well versed in the
chemistry and physics of radicals. I start by attempting a definition of the
term, followed by a very brief backward view, and an outline of the
technique of electron spin resonance (ESR), which is uniquely relevant to
the study of radicals.

My aim is to present an overview that, I hope, enlightens subsequent chapters. Obviously, it is not in any sense exhaustive. I include an
outline of ESR spectroscopy because of its central importance, and because it has been one of my major interests over the past 30 yr.

1

1.1. What Is a Free Radical?

The term "free" has only historical significance and can be dropped with impunity. A broad definition of a radical is that it is a molecule or ion containing an unpaired electron. The significance of this can be gauged in a simple way by considering reaction [1]. Here R· and R·' are

$$R \cdot \ + \ R \cdot ' \ \leftrightarrows \ R\text{–}R' \qquad\qquad [1]$$

radicals, the dots signifying unpaired electrons. These have paired in the bond between the groups in R-R', which is a normal molecule. Whereas most radicals are reactive and undergo dimerization, or other reactions, in which the unpaired electrons become paired, some are relatively stable and have long life-times. These include nitroxide radicals, R_2NO, and a range of radicals and radical-ions in which the unpaired electron is so delocalized that it is unwilling to participate in a localized electron-pair bond. Thus, in a sense, the reactivity of a radical lies in the desire of its characteristic unpaired electron to participate in covalent (electron-pair) bonding. Other reactions of great importance that may precede reactions of type [1] include [2] and [3] [Scheme I], that usually generate new radicals that will ultimately be "destroyed" via reaction [1]. Sometimes, some of these new species have long lifetimes and become readily detectable, especially by ESR spectroscopy (*see under* "spin-trapping, Section 2.4), at others, [2] or [3] may repeat via reactions of R·' so that one initial radical triggers or initiates a series of radical steps before reaction [1] occurs. This is called a *chain reaction*, and is highly characteristic of the reaction of radicals. The very important *autoxidation* reaction, discussed elsewhere, is a key biological reaction that owes its importance to its *chain* nature.

1.2. ESR Spectroscopy: A Unique Tool

Clearly, unpaired electrons (broken bonds) are the key to radical chemistry. It is therefore helpful to consider some properties thereof. Electron spin resonance (ESR) or electron paramagnetic resonance (EPR) (these are the same thing) are spectroscopic techniques that literally involve unpaired electrons and nothing else. So, if a spectrum is obtained, unpaired electrons *must* be present. The converse, unfortunately, is not true as a generalization (*see below*) although a very large range of radicals *can* be ruled out as major reactants if no signal is obtained. ESR spectroscopy is extremely sensitive, and if the spectra comprise narrow lines, concentrations as low as 10^{-7} gm rads λ^{-1} can be detected. A brief overview of the way ESR works and of how it can be used to identify and estimate various radicals is given in Section 2c.

Scheme I
Some Reaction of Radicals

$$R \cdot + R \cdot ' \; \leftrightarrows \; R–R' \qquad\qquad [1]$$

$$R \cdot + >C=C< \; \rightarrow \; \overset{R}{\underset{(R \cdot ')}{C–\dot{C}<}} \qquad\qquad [2]$$

$$R \cdot + R'–H \; \rightarrow \; R–H + R \cdot ' \qquad\qquad [3]$$

R· and R·' are any two types of reactive radical. When R· = R·' reaction [1]$_+$ is a dimerisation. The reverse [1]$_-$ is bond homolysis and may be induced thermally or photochemically. Reaction [2] represents addition to an alkene derivative: depending on the balance of rate constants, etc., [2] can repeat, with R·' adding to another alkene. This important chain reaction is a polymerisation.

1.3. Other Systems Containing Unpaired Electrons

My personal preference is to confine the term "radical" to molecules or ions with one unpaired electron (called *doublet* states because the electron has two quantum numbers, $\pm 1/2$). In the nonmetal field, the most common paramagnetic species other than radicals are those with two unpaired electrons, called *triplet*-states (magnetic quantum numbers $0, \pm 1$). Quite the most important triplet-state molecule is dioxygen, and it is a great pity that ESR spectroscopy can, for various reasons, only be used to detect O_2 in the gas-phase or certain crystalline solids. Other important triplet-states are sometimes obtained on photoexcitation of ordinary (singlet-state) molecules or ions, and these have reactions in some ways typical of diradicals (i.e., species with two radical centers in the molecule). Carbenes and nitrenes are examples of highly reactive groundstate triplet species but, apart from O_2, these species are not discussed herein.

The other important classes of molecules/ions containing more than one unpaired electron are transition-metal and lanthanide ion-complexes. In general (though not always), these do not exhibit the reactions of Scheme I and, hence, are not usually classed as radicals. Often they are very stable, examples being Mn(II) and Ni(II); the major reaction linking them with radicals is that of electron-transfer. (Transition-metal complexes can have up to 5(d) unpaired electrons, whereas lanthanide complexes can have up to 7(f) unpaired electrons.)

Scheme II
Generalized Electron-Transfer Reactions Involving Molecules A
(Electron-Donor) and B (Electron-Acceptor)

$$A \rightarrow A\cdot^+ \quad\quad + e^- \quad\quad\quad [4]$$
(radical-cation)

$$B + e^- \rightarrow B\cdot^- \quad\quad\quad\quad\quad [5]$$
(radical-anion)

$$A + B \rightleftharpoons A\cdot^+ + B\cdot^- \quad\quad\quad\quad [6]$$

(Typical electron-transfer <u>not</u> proceeding <u>via</u> 'free' electrons)

[Note that if A and B and NOT radicals, the products, A·⁺ and

B·⁻ <u>must</u> be.]

1.4. Electron-Transfer Reactions

Molecules, radicals, or ions can participate in electron-transfer reactions (Scheme II). These are unique processes in that they involve movement of a fundamental particle (e^-). (The only other such process in chemistry is proton-transfer, but this is not our concern since electron spin-states are not changed.)

If a singlet-state molecule or ion gains or loses one electron, a radical must be formed (Scheme II). Commonly, e^--gain gives a radical-anion [5], and e^--loss a radical-cation [4].

Some sources of electrons are negative electrodes, anions or molecules with low ionization-potentials, certain photoexcited (triplet-state) species, ionizing radiation and electron-rich transition-metal complexes. Good electron-acceptors include positive electrodes, cations or molecules with high electron-affinities, certain triplet-states, ionizing radiation, and electron-poor transition-metal complexes. Some reactions are shown in Scheme II.

1.5. Aims

My aims in the following sections are:

1. To survey methods of establishing the presence of radicals and their importance in various reactions of biological significance, with special reference to ESR spectroscopy.
2. To show how important oxygen-centered radicals can be, including the role of transition-metal complexes in their formation.
3. Discussion about certain other inorganic and organic radicals (somewhat at random).

4. Finally, discussion about the importance of electron-transfer processes.

2. HOW IS RADICAL INVOLVEMENT ESTABLISHED?

2.1. Chemical Evidence

In many cases, it is clear that a given reaction proceeds via a radical mechanism. Thus, most photochemical reactions are radical in nature, although there are exceptions. Chain reactions are characteristic of radical processes, and are revealed by kinetic studies. Also, reactions that are inhibited by typical radical scavengers are diagnosed as proceeding by a radical mechanism.

Many aspects of these and other approaches are discussed elsewhere in this Chapter, and in various other chapters.

2.2. Detection of Radical Intermediates

This may also be chemical or physical, or a mixture of both. For example, a dramatic method for intercepting radicals is to add a monomer, such as, acrylonitrile ($CH_2 = CHCN$) to the system. When radicals are generated, polymerization ensues and the high polymer is formed as an opaque mass throughout the solution.

However, more direct spectroscopic methods are widely used. Because of its unique ability to detect radicals, and its central role, I concentrate here only on ESR methods. The outline is confined to conventional systems since these dominate current usage. However, highly sophisticated pulse and echo methods and double resonance methods are sometimes, though by no means always, helpful.

2.3. Electron Spin Resonance (ESR) Spectroscopy

This technique is based on the use of homogeneous static magnetic fields and electromagnetic radiation, commonly in the 9000 mHz region (X-band) though lower and higher frequencies are sometimes used and, for some studies, it is helpful to use two quite different frequencies. It is worth stressing the links between this technique and the far better known technique of nuclear magnetic resonance (NMR) spectroscopy. Thus, both the electron (Fig. 1) and the proton (for example) behave as small magnets in a static field, but are quantized in two states ($\pm 1/2$), which can be pictured as the magnetic spins being aligned with, or against, the field. The way these levels diverge as the field increases is shown in Fig. 2, but the rate of divergence is ca. 1000 times faster for the electron than the proton. Resonance occurs when the energy gap between the levels equals the energy of the microwave radiation. A net absorption gives rise to the spectrum. Full details can be obtained, for example, from refs. *1–3*.

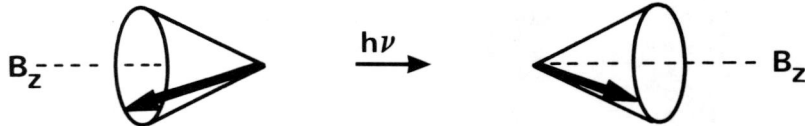

Fig. 1.   The ESR transition induced by the microwave field (hν). The electron is pictured as a small magnet (arrow) precessing about the direction of the
applied magnetic field B_z.

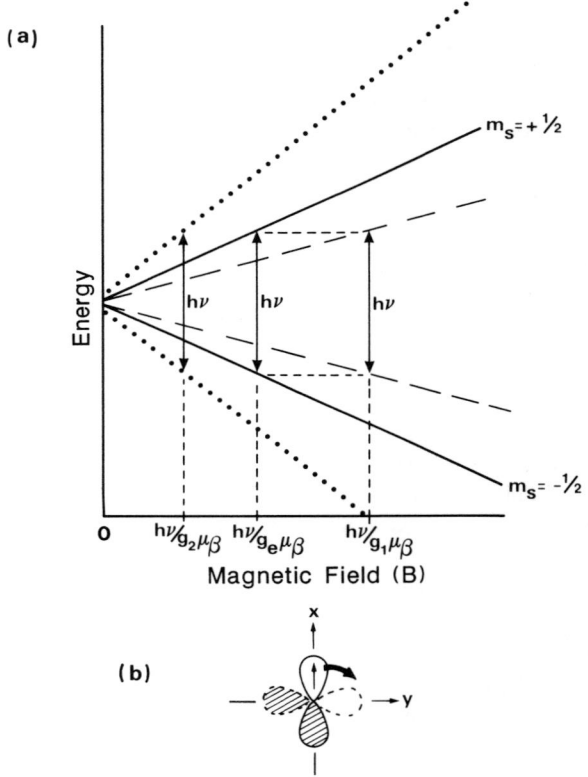

Fig. 2.   (a) The effect of an applied static magnetic field (B) on the $M_S =$
$\pm\frac{1}{2}$ levels of an unpaired electron. These are degenerate at zero field. The thick
lines represent the spin-only behavior, the transition (hν) occurring at a field
corresponding to $g_e = 2.0023$. The dashed lines are for a radical having a low-
lying vacant excited state, giving $g_1 < 2.0023$, and the dotted lines are for a
radical having a neighboring filled level, giving $g_2 > 2.0023$ [taken from Symons
1978 (1)]. (b) Movement of the unpaired electron from one p-orbital (p_x) into
another (p_y), induced by a magnetic field (B_z) gives rise to orbital motion and,
hence, a shift in the g-value.

2.3.1. The g-Value

In ESR spectroscopy, it is normal to use a fixed microwave frequency ($h\nu$), and to vary the magnetic field (B). The rate of divergence of the levels (g) is then given by the first equation in Scheme III.

2.3.2. Electron-Nuclear Hyperfine Coupling

If there are magnetic nuclei present, these may couple with the electron to give a characteristic set of features. This can be compared with spin–spin coupling in NMR and, for liquid phase spectra, spectral analysis is comparable (*see* Scheme III and Figs. 3–7).

Figure 3 is the same as Fig. 2, but includes the effect of hyperfine coupling to a single nucleus of spin ½. So for each electron level, there are two nuclear levels, giving four in all. Generally, when the electron is at resonance and undergoes a 'spin-flip,' the nucleus is unaffected. The trends are shown only for the linear, high-field, limit. They curve at low fields, but this is only important for very large hyperfine coupling constants (*1,28,30*).

Figure 4 shows the doublet spectrum for H· atoms and the triplet for D· atoms (^2H has $I = 1$, $M_I = \pm 1,0$). Note the marked change on going from H· to D·. This decrease in hyperfine coupling is frequently used to identify specific protons in radicals, or to reduce the total splitting so that broad lines are narrowed.

The situation for a typical nitroxide radical is illustrated in Fig. 5. This could be $(Me_3C)_2NO$ with resolved coupling only to ^{14}N nuclei ($I = 1$), giving a characteristic triplet (a). If this system is frozen to a glass, the spectrum becomes much broader but, in contrast with NMR spectra, can still be extremely useful. Whereas not as informative as a single crystal study, maximum and minimum hyperfine splittings can usually be measured, as shown, and these can be used to characterize the radical. Note that anisotropy in the hyperfine coupling and g-value is a cause of line broadening in the liquid-phase. In general, the greater the field range covered by a feature in the solid-state, the broader the line in fluid solutions, unless the viscosity is low. If radicals are part of a biopolymer that tumbles slowly, ESR features may be greatly broadened by this effect.

Hyperfine coupling can be used as a "fingerprint" to identify radicals (as in NMR) but it also gives considerable structural information, especially if anisotropic parameters are available. Indeed, this has been an important source of structural information for many organic and inorganic radicals. This work has been an essential background to the application of ESR methods to biological systems.

The way in which electrons couple to nuclei magnetically to give hyperfine splitting is illustrated in Figs. 3–5. For a more detailed, non-mathematical explanation, see ref. *30*, and for in-depth quantum mechanical discussion, *see* ref. *1* or *28*.

Scheme III
Some ESR Relationships

$h\nu = g\ \mu_\beta\ B$

$h\nu = g\ \mu_\beta\ (B + M_I\ A)$

$g_e = 2.0023$ (for zero orbital angular momentum – the
        "free-spin" value)

$\Delta g = (g_{exp} - g_e)$

Δg is a measure of induced orbital magnetism; it can be compared
    with the chemical shift in n.m.r.

ν = microwave frequency (often ca. 3000 MHz; X-band)

A = hyperfine coupling (between the electron and any magnetic
    nuclei. Thus, if $I = \frac{1}{2}$ (i.e. a proton) $h\nu = g\ \mu_\beta\ (B \pm \frac{1}{2}A)$
    and two lines separated by A result, in the high field limit.

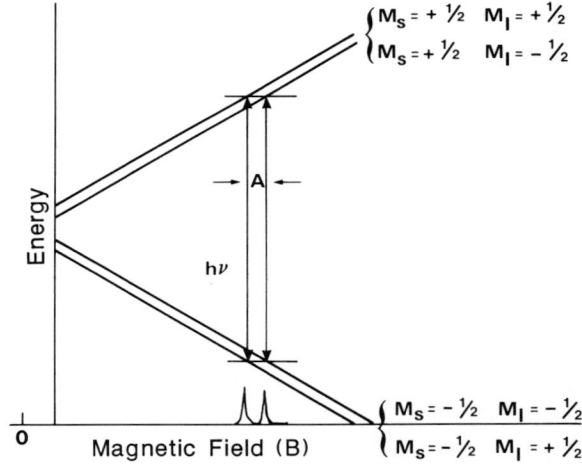

Fig. 3. Divergence with field of the $M_S = \pm 1/2$ levels in the presence of a single nucleus having $I = 1/2$, in the high-field approximation. [Note the two allowed transitions involve no change in M_I.]

2.4. Spin-Trapping

This is a general technique in which radicals, that are not readily observable by ESR directly, react with a radical-scavenger or "spin-trap" to give a new radical having a readily detectable spectrum and, in general, a much longer life-time. By far the most popular spin-traps are nitroxide radical precursors, since many nitroxides are long-lived and have characteristic ESR spectra (*see* Fig. 5 and Scheme IV).

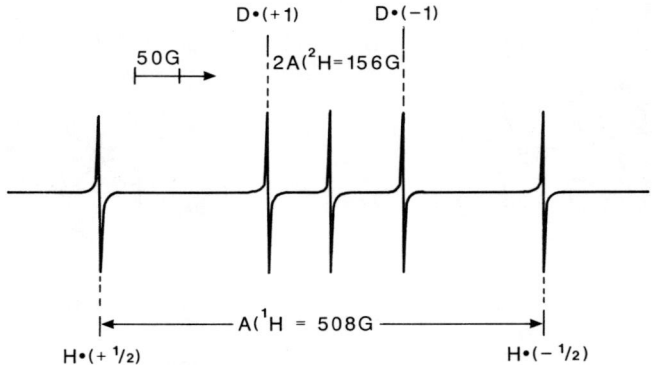

Fig. 4.   First derivative ESR spectrum showing features for trapped hydrogen and deuterium atoms.

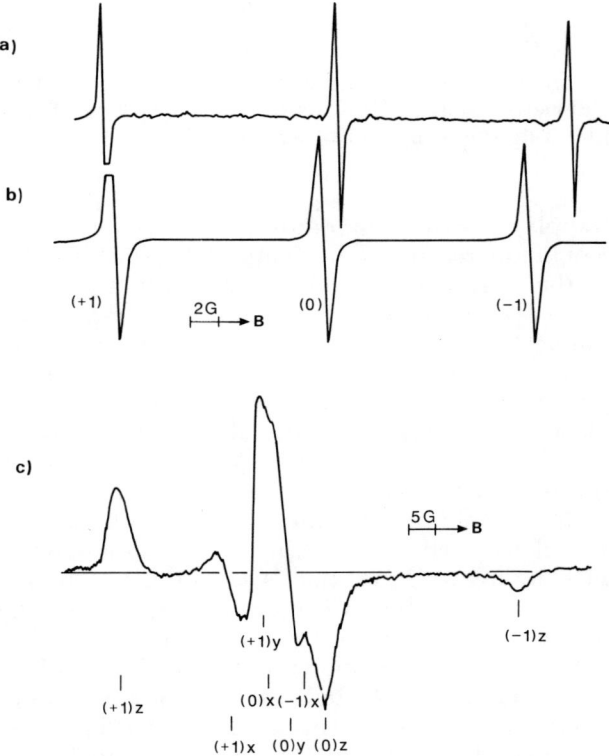

Fig. 5.   First derivative X-band ESR spectra for $(Me_3C)_2NO\cdot$ radicals (a) in water at 20°C; (b) in dodecane at 20°C, and (c) in CD_3OD at 77 K. In (c), $[(CO_3)_3]_2NO\cdot$ radicals were used to give better definition.

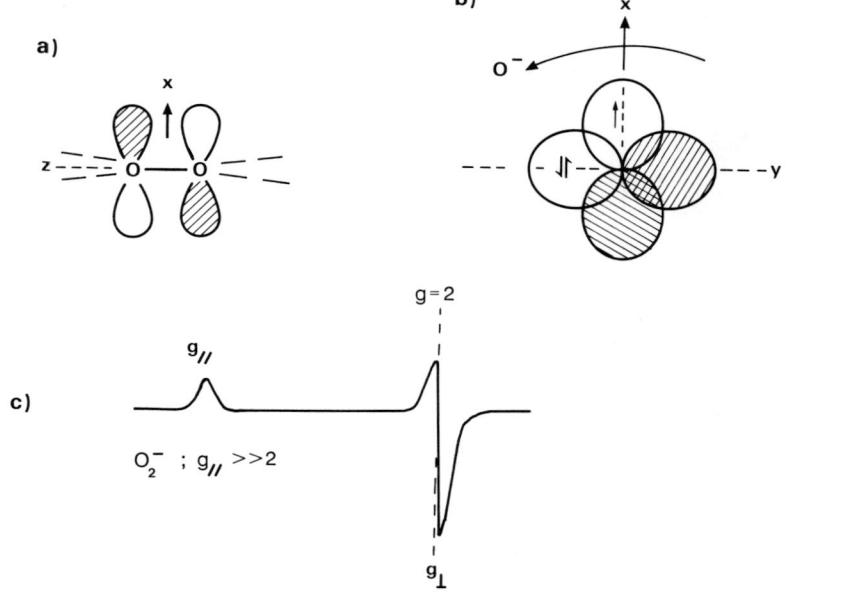

Fig. 6.   The O_2^- radical: (a) the π_x^* orbital containing the unpaired electron; (b) as (a) viewed along z, showing the filled π_y^* orbital. The dashed lines in (a) and (b) indicate the formation of hydrogen bonds by solvent molecules.

Spin-trapping has two rather different roles. The most obvious is the conversion of highly reactive radicals into stable (persistent) radicals, and the other is the conversion of radicals that cannot be detected by ESR normally, into species that can. As discussed below, unfortunately, many very important radicals involved in biochemical processes, such as, ·OH or ·O_2^- are undetectable by ESR in the liquid state, so the only thing to do is to convert them into stable species.

There is a loss, however, because all nitroxide radicals have rather similar spectra, so it is often difficult to identify the parent radical from its nitroxide spectrum. The best situation is when *extra* hyperfine coupling is resolved. (If this cannot be seen directly, resolution may be achieved by ENDOR and related methods (1,28,30).) A nice example is the detection of hyperfine coupling to ^{13}C for the [13]·CCl$_3$ adduct of the spin-trap, PBN (*see* Table 1 and Fig. 7). The ·CCl$_3$ radicals were not directly detectable.

Failing that, changes in the coupling constant to ^{14}N or ^{1}H of the nitroxides themselves can be monitored. The ^{14}N coupling is not very sensitive to the nature of R·, but, for the nitrone spin-traps, the ^{1}H coupling can be very sensitive. This arises because of its conformational dependence outlined in Scheme IV. There is no way of predicting the ^{1}H coupling for a given adduct except that bulky groups will tend to occupy out-of-plane sites, thereby constraining the hydrogen towards the radical plane, and hence, a small splitting. What is done is to generate various

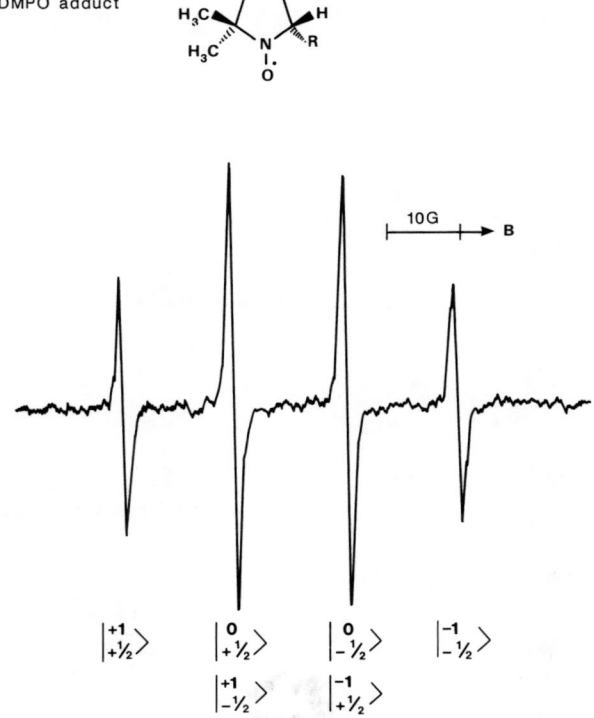

DMPO adduct

Fig. 7. First derivative X-band ESR spectrum assigned to the ·OH radical adduct of DMPO. [Note that coincidentally, $A(^1H) = A(^{14}N)$ so that the spectrum is a quartet (1:2:2:1).]

radicals by unambiguous methods, and to study their spin-trap adducts. These spectra are then taken to be characteristic of the radicals in question. Whereas not being infallible, this method seems to be very useful in practice.

3. SOME INORGANIC RADICALS

3.1. Dioxygen in Triplet and Singlet States

A major source of radicals in biological systems is dioxygen (O_2). This very reactive molecule, although essential to the life of higher organisms is, nevertheless, very dangerous in excess or, for example, in anoxia-reperfusion situations.

Dioxygen has a ground-state triplet level, the two unpaired electrons being accommodated, formally, in the degenerate pair of antibonding π-orbitals, $\pi_x^*(\uparrow), \pi_y^*(\uparrow)$. It is noteworthy that, because of strong coup-

Scheme IV
Aspects of Spin-Trapping and Examples of Commonly Used Spin-Traps

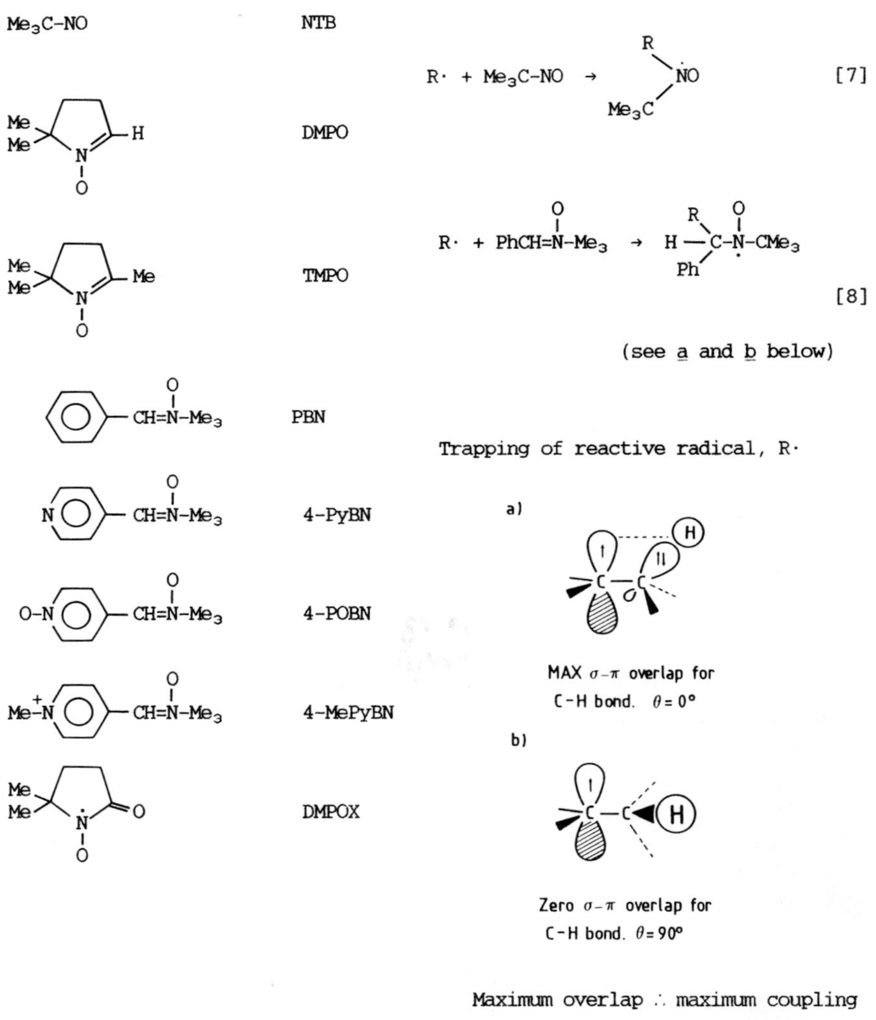

Me_3C-NO NTB

DMPO

$$R\cdot \; + \; Me_3C-NO \quad \rightarrow \quad \underset{Me_3C}{\overset{R}{\diagdown}} \dot{N}O \qquad [7]$$

TMPO

$$R\cdot \; + \; PhCH=\overset{O}{\overset{|}{N}}-Me_3 \quad \rightarrow \quad \underset{Ph}{\overset{R}{H-\overset{|}{C}}}-\overset{O}{\overset{|}{N}}-CMe_3$$

[8]

(see a and b below)

PBN

Trapping of reactive radical, R·

4-PyBN

a)

4-POBN

MAX $\sigma-\pi$ overlap for
C–H bond. $\theta = 0°$

4-MePyBN

b)

DMPOX

Zero $\sigma-\pi$ overlap for
C–H bond. $\theta = 90°$

Maximum overlap ∴ maximum coupling

Inserts a and b show why the ^{1}H hyperfine coupling is very sensitive to the size and nature of the added radical, R. a The electron spin can delocalize onto the proton giving a large hyperfine coupling. b No delocalization is possible for this conformation.

Table 1
Typical ESR Parameters for Various Inorganic Radicals

Radical	Matrix	g Values				Nucleus	Hyperfine Coupling Constants/G[a]				Reference
		g_x	g_y	g_z	g_{av}		A_x	A_y	A_z	A_{iso}	
$O^{-}\cdot$	D_2O/NaOD	2.070	2.002	2.002	2.025	—	—	—	—	—	6
$OH\cdot$	H_2O	2.059	2.009	2.0027	—	^1H	0	45	28.5	24.5	8
$O_2^{-}\cdot$	MeOH ± H_2O	2.078	2.000	2.000	2.026	—	—	—	—	—	32
$HO_2\cdot$	H_2O_2 ± H_2O	2.0353	2.0086	2.0042	2.0160	^1H	13.8	3.6	15.7	11.0	28
$O_3^{-}\cdot$	$KClO_3$	2.0025	2.0174	2.0113	2.0104	—	—	—	—	—	14
$Cl_2^{-}\cdot$	H_2O	2.000	2.040	2.040	2.027	^{35}Cl	102	10	10	40.8	9, 10
$ClOH^{-}\cdot$	H_2O^g	2.004	2.017	2.017	2.013	^{35}Cl	59	−16	−16	9	15
$ClOO\cdot$	$KClO_3$	1.9983	2.0017	2.0130	2.0043	^{35}Cl	5.3	7.2	14.9	9.1	14
$ClO\cdot$	CO_2	1.889	1.899	2.66	2.149	^{35}Cl	30	ca. 0	ca. 10	ca. 10	35
$\cdot NO_2$	H_2O	2.0066	1.9920	2.0022	2.0003	^{14}N	50.6	49.84	70.21	56.88	2
$\cdot SO_2^{-}$	H_2O	—	—	—	2.0057	^{33}S	—	—	—	14.2	3
		—	—	2.0026	2.0026	^{32}S	153	115.5	115.5	128	25
$\cdot SO_3^{-}$	H_2O	—	—	—	2.0028						

[a] 1 G = 10^{-4} T.

ling to rotational levels, the ESR spectrum for O_2 as a low-pressure gas comprises sets of many narrow lines spread over a wide field range. Unfortunately, these are so extensively broadened for O_2 in solution that no resonance is detectable. However, oxygen is an important source of internal fluctuating magnetic fields, that may broaden the ESR features of other radicals. It is important to note that there are several low-lying excited states for dioxygen, probably the most important being the $^1\Delta$ state in which the two electrons are, formally, paired in one of the π^*-orbitals, leaving the other vacant. The $(\pi_x^*)^2,(\pi_y^*)^0$ description is not strictly correct for the gas-phase molecule, but I suggest that it may be suitable for $^1\Delta O_2$ in aqueous solution. This description stems from our knowledge of the asymmetric solvation of O_2^-, in which hydrogen bonding is directed towards the filled π^*-orbital only (cf Fig. 6). I think some elements of this solvation might also occur for aqueous solutions of the $^1\Delta$ form of dioxygen. This would serve to distinguish between π_x and π_y. It might also have a measurable effect on the emission spectrum for $^1\Delta$ as it falls to the ground state. Such an effect could be important because this spectrum is used for identification of singlet dioxygen. (*17,18*).

Dioxygen in its ground (triplet) state reacts as a rather stable diradical. One of its most important reactions is [9], in which radicals R· are converted into peroxy radicals R-OO· that generally have quite different reactivities from those of the parent R· species. This reaction is, in general, reversible, and stable R· species may fail to undergo reaction [9] to any measurable extent.

$$R \cdot + {}^3O_2 \rightleftharpoons R\dot{O}O \qquad [9]$$

In contrast, $^1\Delta$ oxygen, often just referred to as 'singlet oxygen,' reacts more as an electrophile, via the "empty" π^* orbital. Thus, its reactions and reactivity are rather different. Of course, being an excited state, it has a limited life-time, and in the absence of reaction, it falls to the ground state with light emission (in the 1270 nm region for spontaneous, unimolecular emission (*17,18*). In the absence of collisions (e.g., in the upper atmosphere), its life-time may be ca. 1h, (*15*) but in solution, the quantum efficiency of light emission falls drastically, even though its rate is greatly increased (*27*). Probably, solvent complexes are formed, which facilitate radiationless decay and light emission.

Singlet oxygen is formed from excited-states of various sensitizers, such as, acridine. It is thought to be of importance in various biological contexts, one important example being "photodynamic cancer therapy." I stress, however, that it is normal triplet oxygen that has radical characteristics rather than singlet oxygen.

3.2. O_2^-, $HO_2\cdot$, and $RO_2\cdot$

The pK_a for $HO_2\cdot$ is ca. 4.9 in water. Hence at pH 7, a small but significant amount of O_2^- will be protonated. In acidic regions of the body, such protonation may be very important and, since certain cancer cells are thought to have a relatively low pH, $HO_2\cdot$ may be of greater importance therein.

The *superoxide ion* has its unpaired electron in a π^* orbital (Fig. 6). There is no distinction between π_x^* and π_y^* in the absence of solvation but, in our view (32,33), strong hydrogen-bonding occurs primarily on one plane, as indicated in Fig. 6. This will favor the filled π^* orbital, forcing the unpaired electron into the "unsolvated" orbital. It seems that this is a major phenomenon, since $\Delta g_{\parallel}$, which increases as the π_x^*/π_y^* splitting falls, is, in fact, quite small for frozen aqueous solutions (ca. + 0.07). I stress that the magnitude of the downfield shift for $g_{\parallel}$ is not a characteristic datum for O_2^- as such, but is a property of the electric-field (hydrogen-bonding) that controls Δg.

No ESR signal has even been observed for O_2^- in liquid water. This is expected because of the relatively large field modulation caused by the rotating ion, and the uncertainty in g_{av} caused by fluctuations in the degree and precision of hydrogen-bonding. The result is a line that is too broad to be detected.

Salts of the superoxide ion, such as KO_2, are commercially available, and provide a ready source of O_2^- in solution. Alternatively, it can be formed by ionization of $HO_2\cdot$ (readily formed from H_2O_2), or by electron-addition to O_2 using, for example, ionizing radiation.

It is important to note that O_2^- is a weak base, as well as being a radical. Its basicity, or nucleophilicity, is strongly a function of solvation, as for all anionic bases, increasing enormously as the extent of hydrogen-bonding is reduced. This has, at least, two implications: (i) on initial formation from O_2, O_2^- will be a far stronger base than it is once fully hydrated; (ii) when O_2^- is formed from O_2 in a nonaqueous environment, such as a lipid membrane, or a hydrophobic pocket of a protein, it will be a strong base, and this could be its overriding mode of reaction in view of its low reactivity as a radical (24,26).

Two aspects of our own work illustrate these points. In one (33), O_2^- was formed by electron addition to O_2 in glassy ethanol at 4 K. Dioxygen does not form hydrogen bonds in ethanol, and hence, the O_2^- is initially unsolvated. Solvent–solvent hydrogen-bonds have to break before solvation can occur, and the complete absence of any ESR spectrum assignable to O_2^- shows that it remains unsolvated at 4 K. However, on warming, a broad parallel feature grew in, and this shifted toward the free-spin region and narrowed markedly on further heating. These changes occurred in stages and, in all, four different spectra could be discerned. The final spectrum, obtained on warming to ca. 90 K (well below the softening temperature) was identical with that for fully solvated O_2^-.

The other study showed that generation of $O_2^{\cdot-}$ from O_2 in dry dimethylformamide gave a species whose ESR spectrum agrees closely with that expected for an $RO_2\cdot$ radical, not $O_2^{\cdot-}$, which would probably remain undetected in this solvent (7). We suggest that reaction [10], comprising *nucleophilic* addition to the carbonyl group, is responsible.

$$O_2^{\cdot-} + \quad \underset{Me_2N}{\overset{H}{>}}C{=}O \quad \leftrightharpoons \quad \underset{Me_2N}{\overset{H}{>}}C\underset{O^-}{\overset{O\dot{O}}{<}} \qquad [10]$$

Addition of water readily reverses this reaction. However, for esters, the reaction proceeds as in [11], and it has been suggested that this could be an important reaction for $O_2^{\cdot-}$ generated in lipid membranes (10,11). Both the intermediate and the acylperoxy radical should be more reactive as *radicals* than $O_2^{\cdot-}$.

$$O_2^{\cdot-} + \quad \underset{R'O}{\overset{R}{>}}C{=}O \quad \leftrightharpoons \quad \underset{R'O}{\overset{R}{>}}C\underset{O^-}{\overset{O\dot{O}}{<}} \quad \rightarrow \quad RC\underset{O}{\overset{O\dot{O}}{<}} + R'O^- \; [11]$$

These considerations are important because of the high solubility of oxygen in nonaqueous solvents relative to that in water. Of course, since $O_2^{\cdot-}$ is greatly stabilized by water, it will be a far better *electron*-donor in aprotic solvents than it is in water. Thus, in reaction [12] where A is,

$$O_2^{\cdot-} + A \quad \leftrightharpoons \quad O_2 + A^{\cdot-} \qquad [12]$$

say, a large aromatic compound, the solvation of $O_2^{\cdot-}$ will be far greater in water than that of $A^{\cdot-}$. Hence, equilibrium [12] could strongly favor $O_2^{\cdot-}$ in water, but $A^{\cdot-}$ in a lipid environment.

The ESR spectra for $HO_2\cdot$ and $RO_2\cdot$ radicals are very similar (cf Fig. 7) except that the former shows hyperfine coupling to the proton (9,10), whereas $RO_2\cdot$ radicals generally show no ^1H coupling. The much smaller g-shift relative to $O_2^{\cdot-}$ arises because addition of H or R lifts the orbital degeneracy very strongly. In principle, both $HO_2\cdot$ and $RO_2\cdot$ radicals should be detectable in fluid solution but, in practice, only $RO_2\cdot$ radicals have so far been 'seen.'

These radicals are more reactive than $O_2^{\cdot-}$, and can extract 'active' hydrogen as, for example, in reaction [13].

$$RO_2\cdot + R'CH_2CH{=}CHR'' \rightarrow RO_2H + R'\overline{\dot{C}HCHCHR''} \qquad [13]$$

The resulting hydroperoxide, like H_2O_2, is quite reactive, as is the 'allyl' radical that, formally, has ca. 50% spin-density on two almost equivalent carbon atoms, as indicated. This is required by simple theory, and has been nicely established by ESR spectroscopy (19). Reaction [13] is one of major importance in membrane chemistry, since it is one of the stages in the autoxidation of unsaturated lipid groups. The total reaction comprises [13] followed by [14] which, in turn is followed by [13]. These two

$$R'\overline{CH-CH-CH}R'' + O_2 \quad \rightarrow \quad R'C—CH=CHR'' \qquad [14]$$

$$\overset{|}{O_2} \text{ (RO}_2 \cdot)$$

steps constitute a chain reaction that may continue until the supply of oxygen runs out. The key factor is, of course, that one active radical formed in the membrane may lead to many damaged lipid groups. Such chain reactions require a fine balance of reactivities, and are rare in biological systems.

We have now seen two very important roles for dioxygen, namely, electron addition and addition to carbon-centered (and other) radicals to give $RO_2\cdot$ (peroxy) radicals. Both are of great importance in biology, although, generally, biological reactions avoid both, unless the products (O_2^- and $RO_2\cdot$) are needed in some specific process (phagocytosis is a possible example). Thus, enzymes involved in the conversion of dioxygen into water react via two electron addition stages, so that the only reactive intermediate is H_2O_2 (or HO_2^-). Even, this may be avoided in certain cases, the overall "4-electron" reaction to give $2H_2O$ all occurring "safely" within a complex by 'contained' chain of reactions. Possibly, one danger is to overwhelm such enzyme systems such that they are unable to return to their resting forms before another substrate molecule (such as oxygen) is taken up. Hence, the normal 2- or 4-electron stage is unavailable. An example seems to be the formation of O_2^- by xanthine oxidase. This enzyme has the ability to handle multielectron reactions so O_2^- formation is probably anomalous. As is stressed elsewhere, when certain active radicals *are* formed, there are specific enzymes that readily scavenge and decompose them. The superoxide dismutases are a good example, in the catalysis of reaction [15].

$$2O_2^- + 2H_2O \leftrightharpoons H_2O_2 + O_2 + 2OH^- \qquad [15]$$

3.3. The Hydroxyl Radical

This is by far the most reactive of the oxygen centered radicals, but this high reactivity doesn't mean that it is the most significant. From an ESR viewpoint, like O_2^-, it has a degenerate π system and $\pi_x(\rho_x)$ must be different from $\pi_y(\rho_y)$ if an ESR spectrum is to be detected. This splitting is again achieved by hydrogen-bonding, but, as with O_2^- no liquid-phase ESR spectrum can be detected, although in ice, the solid-state spectrum is well defined (8). Under carefully controlled conditions, spin-trap adducts can be formed, the most characteristic being that of DMPO (Fig. 7).

Hydroxyl radicals are formed from water by electron-loss ($H_2O\cdot^+$ is a strong acid and rapidly loses a proton) but this is normally only achieved by ionizing radiation. The most important mode of generation is from H_2O_2, either by bond homolysis (UV-light) or by electron capture. It is generally agreed that certain transition-metal aquoions can achieve this, but complexed ions may well react by alternative routes (*see* Chapter 3).

The key to understanding the importance of ·OH radicals in tissues is to realize that they are *so* reactive that their reaction rates are usually close to the diffusion controlled limit. (They react either by H-atom extraction or by addition to double bonds. Thus, for example, they *add* to DNA bases and extract hydrogen from the deoxyribose units in a competitive fashion.) This means that they react indiscriminately with the first biomolecule they encounter. Thus, damage is widespread but indiscriminate. They will not travel far, and reaction schemes in which, say, H_2O_2 reacts with membrane-bound enzymes to give ·OH radicals that then are thought to damage DNA, are surely invalid. However, systems that generate ·OH close to DNA will be most effective in inducing DNA damage. An example is bleomycin, that binds strongly to DNA and has a high affinity for iron. This can react with H_2O_2 to give ·OH, which then may attack DNA.

The pK_a of ·OH is close to that of water, so O^- ions are unlikely to be of biological importance. The ·OH radical has a high electron-affinity, and its reactions are sometimes described as being "electrophilic" in nature. Thus, it prefers to react at electron-rich sites in molecules, although, in fact, it is not very selective.

I conclude that, although the ·OH radical is the most reactive of these "oxygen radicals," it may not be as dangerous as, say, $HO_2\cdot$, because of the speed and indiscriminate nature of its reactions. Only when it is generated very close to its target is it expected to be important.

3.4. Other Inorganic Radicals

3.4.1. Solvated Electrons

In radiolyses, solvated electrons, usually written as e^-_{aq}, are of great importance (12,29). These unique species, thought to be structurally related to F-centers in alkali-halide crystals (13), readily add to electron-affinic compounds. However, it is doubtful if they are ever important in a normal biological context, and so, they will not be considered herein.

3.4.2. Hydrogen Atoms

Again, these reactive atoms are formed in radiolyses and certain photochemical reactions, but are not thought to be of biological significance.

A list of some other, possibly important, inorganic radicals is given in Table 1, together with some ESR parameters. A discussion of the structure and ESR spectra of these radicals is given in refs. 4, 30, and 31.

4. SOME ORGANIC RADICALS

A very wide variety of organic radicals have now been characterized by their ESR spectra and, in many cases, by UV or IR spectra, as well (*see*, for example, 2b, small radicals (or radical-ions)).

In the solid-state (noncrystalline), these features are usually very broad because the hyperfine coupling is anisotropic, with x, y, and z features. Hence the sensitivity is greatly reduced.

The nitroxides have already been discussed. Another example of an organic radical having a remarkably simple spectrum is the benzosemi-quinone anion (I).

I

4 equal ^{1}H nuclei
A(^{1}H) = 2.37 G

This has four equivalent coupled protons, and hence, gives rise to a 1:4:6:4:1 set of equally spaced lines. There is considerable delocalization of the unpaired electron onto oxygen, and this gives rise to a positive increment to Δg, with $g_{av} = 2.0047$. This sort of shift is a characteristic aspect of this biologically important class of radicals.

For most carbon-centered radicals, the problem of orbital degeneracy discussed for O_2^- and ·OH radicals doesn't arise, and narrow-lined liquid-phase ESR spectra are obtained. Radicals are then fingerprinted by their hyperfine features, these usually being caused by electron–proton coupling. A simple example is that of the methyl radical, ·CH$_3$, characterized by a 1:3:3:1 quartet (Fig. 8). Other alkyl radicals all give well-resolved, narrow features in the liquid-phase.

Unfortunately, many of the radicals expected to be formed from biomolecules will give solid-state type spectra, with broad lines, and hence, detection by ESR spectroscopy is difficult. This occurs because biomolecules are frequently polymeric, and their rate of tumbling is too slow to average out the anisotropy.

Most of the bioradicals that have been studied directly are relatively stable, small radicals. Often, these are of the semiquinone type, in which the unpaired electron is delocalized in the π-system, and has little tendency to react. These species are often formed by electron-gain or -loss from diamagnetic parent molecules. A very important radical, possibly the most frequently detected of all by the ESR technique, is the ascorbate radical, formed by electron-loss from the ascorbate anion, (II), followed by proton loss, or by hydrogen atom transfer. It is, of course, an inter-

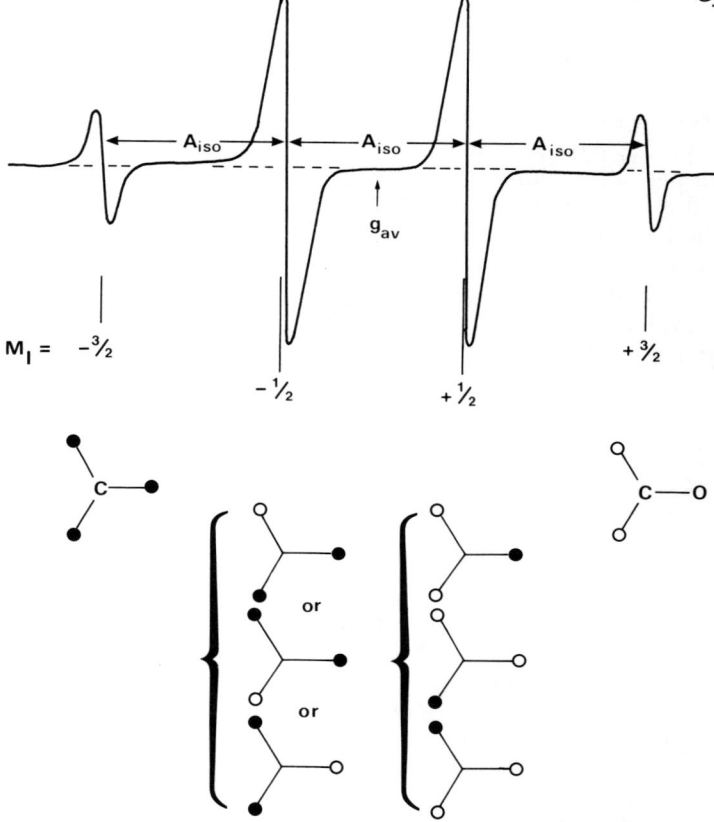

Fig. 8.   First derivative ESR spectrum for methyl radicals in solution. The various combinations of proton spins is indicated to show why the relative intensities are 1:3:3:1. [● represents (−1/2) and ○ (+1/2) spin states.]

mediate in the conversion of ascorbate into dehydroascorbate. Its ESR spectrum is characterized by a doublet splitting of 1.8 G, which makes it easy to identify under high resolution conditions.

(II)

5. SULPHUR-CENTERED RADICALS

Radicals of type RS· are of considerable importance in the field of bioradicals. Although they are structurally similar to ·OH or, more generally, to RO· radicals, they are much less reactive, and hence RSH compounds are good H-atom donors, and RS⁻ ions, readily formed from RSH compounds, are good electron-donors.

Yet again, these important radicals have large, solvent-dependent $g_{\parallel}$ shifts, and hence, cannot be detected by ESR spectroscopy in the liquid-phase. Fortunately, they readily react with the parent RSH (or RS⁻), or indeed, any RSH compounds, to give stable RS∸SR⁻ radicals, that can be detected in the liquid phase, although Δg is still considerable (i.e., g_x ~2.02, g_y ~2.015, and g_z ~2.002) and, hence, the features are relatively broad. They are also characterized by a very intense visible absorption (λ_{max} ~400 nm). Some reactions relating to these radicals are given in Eq. [16]–[19]. The RS∸SR⁻ molecules are typical of a class of molecule having a "three-electron" bond. This means that, in addition to the normal pair of electrons in the σ-bonding orbital, there is an extra electron in the σ*, antibonding, orbital.

$$RS^- \longrightarrow RS\cdot + e^- \qquad [16]$$

$$RS\cdot + RS^- \rightleftharpoons RS\dot{\bar{}}SR^- \qquad [17]$$

$$RS-SR + e^- \longrightarrow RS\dot{\bar{}}SR^- \qquad [18]$$

$$RS\cdot + R_2S \rightleftharpoons RS\dot{\bar{}}SR_2 \qquad [19]$$

The situation is summarized in Fig. 9 for the generalized A∸B⁻ radical. For completeness, I include the structure for σ^1 radicals (having a one-electron σ-orbital), although these are not thought to be formed in biological systems.

It is important to note that RO· radicals do *not* normally react to give RO∸OR⁻ or related σ* species. Also, RO–OR molecules, on electron capture, dissociate to give RO· and RO⁻ rather than RO∸OR⁻.

RSH compounds, such as, cysteine and glutathione, are thought to be of importance in rapid repair of radical damage, especially to DNA (22,23). Ascorbate may play a similar role.

6. NITROXIDE RADICALS

These radicals, R_2NO (isoelectronic with ketyl radical anions R_2CO^-) are by far the most studied radicals using the ESR technique (*see above*). When the R-groups are bulky, as for $(Me_3C)_2NO$, they show no tendency to dimerize and have very low radical reactivity. It is surprising to know that they do not occur as such in biological systems. Nevertheless, they are of outstanding importance in the study of bioradicals, because reac-

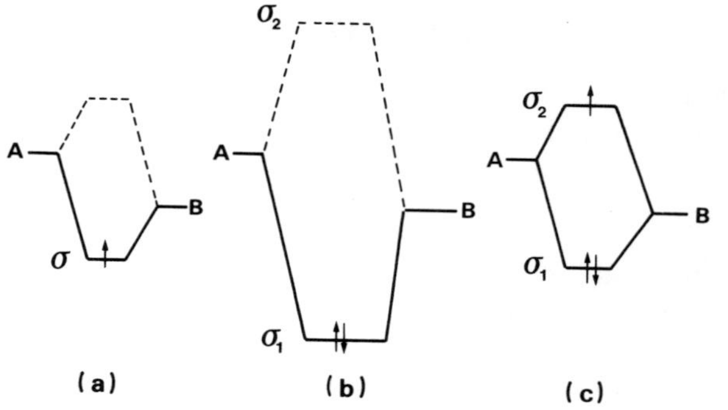

$$\sigma_2 \qquad \sigma_2$$

(a)   (b)   (c)

Examples of $\sigma*$ Radicals (c)

$$Cl \dot{-} Cl^- \qquad R-Cl \dot{-} Cl-R^+$$

$$RS \dot{-} SR^- \qquad R_2S \dot{-} SR_2^+$$

$$R_3P \dot{-} PR_3^+ \qquad R_2S \dot{-} Cl$$

Fig. 9. Qualitative energy level diagram for $\sigma_1{}^1$ cations (a) and $\sigma_1{}^2\sigma_2{}^1$ anion-radicals (c) relative to that for the parent molecule (b).

tive, or undetectable, radicals can often be converted into nitroxides. Two classes of precursors are used, as indicated in Scheme IV.

Using one of these "spin-traps," many small, reactive, bioradicals have been studied. The extensive work of Mason and coworkers illustrates this method for direct studies of radicals in tissues (21–23). An early and convincing study is that of Slater and coworkers on the $\cdot CCl_3$ radical, formed from CCl_4 in the liver. Using ^{13}C labeled material, a nitroxide ESR signal with a well-defined ^{13}C splitting was obtained (34). This is one of the most convincing studies showing that radicals are indeed important intermediates in metabolism. An intriguing complementary study has shown that $^{13}\cdot CCl_3$ radicals go on to form the hydrolysis product, $^{13}\cdot CO_2{}^-$, that was also characterized by spin-trapping.

7. CONCLUSIONS

It is clear that radicals are often important intermediates in biological reactions. However, *proving* their involvement is a difficult task. This is exacerbated by the fact that some of the most important radicals ($\cdot OH$, $\cdot O_2{}^-$, $HO_2\cdot$, $RO_2\cdot$) cannot normally be detected directly, and one relies on various indirect tests that, in some cases, are not able to identify the intermediates with clear precision.

Much more work is needed!

REFERENCES

1. N. M. Atherton, *Electron Spin Resonance*, Halsted, London, 1973.
2. P. W. Atkins, J. A. Brivati, M. Keen, M. C. R. Symons, and P. A. Trevalion. Oxides and oxyions of the nonmetals: CO_2^- and NO_2. *J. Chem. Soc.* 5220–5225 (1962).
3. P. W. Atkins, A. Horsfield, and M. C. R. Symons. Oxides and oxyions of the nonmetals: SO_2^- and ClO_2. *J. Chem. Soc.* 5220–5225, (1964).
4. P. W. Atkins and M. C. R. Symons. *The Structure of Inorganic Radicals*, Elsevier, Amsterdam, 1967.
5. R. M. Badger, A. C. Wright, and R. F. Whittock. Absolute intensities of the discrete and continuous absorption band of oxygen gas at 1.26 and 1.065 μ and the radiative lifetime of the singlet state of oxygen. *J. Chem. Phys.* **43**, 4345–4350 (1965).
6. M. J. Blandamer, L. Shields, and M. C. R. Symons. Solvated electrons: Stabilization in aqueous alkali-metal hydroxide glasses. *J. Chem. Soc.* 4352–4357 (1964).
7. P. J. Boon, M. T. Olm, and M. C. R. Symons. Electron spin resonance and ENDOR spectra of radicals formed by the addition of superoxide ions to dimethylformamide. *J. Chem. Soc. Faraday Trans.* **84**, 3334–3345 (1988).
8. J. A. Brivati, M. C. R. Symons, D. J. A. Tinling, H. W. Wordale, and D. O. Williams. Electron spin resonance studies of the hydroxyl radical in γ-irradiated ice. *Trans. Faraday Soc.* **63**, 2112–2119 (1967).
9. R. C. Catton and M. C. R. Symons. Unstable intermediates. Part L.VIII An electron spin resonance study of γ-irradiated Group 11A chloride hydrates. *J. Chem. Soc.* **A**, 446–451, (1969).
10. R. C. Catton and M. C. R. Symons. Unstable intermediates. Part LX. HO_2 radical in γ-irradiated strontium chloride hexahydrate. *J. Chem. Soc.* **A**, 1393–1395 (1969).
11. P. M. Cullis, G. D. D. Jones, J. S. Lea, M. C. R. Symons, and M. C. Sweeney. The effect of ionizing radiation on DNA: The role of thiols in chemical repair. *J. Chem. Soc. Perkin Trans.* **2**, 1907–1914 (1987).
12. F. W. Dainton. The electron as a chemical entity, *Faraday Lecture, Chem. Soc. Rev.* **4**, 323–362 (1975).
13. C. J. Delbecq. A study of M center formation in additively colored KCl. *Z. Phys.* **171**, 560–566 (1966).
14. R. S. Eachus, P.R. Edwards, S. Subramanian, and M. C. R. Symons. Oxides and oxyanions of the nonmetals. Part VIII. Electron spin resonance studies of some paramagnetic chloride oxides. *J. Chem. Soc.* **A**, 1704–1711 (1968).
15. I. S. Ginns and M. C. R. Symons. Unstable intermediates Part CV. Radiolysis of frozen aqueous solutions of alkali-metal halides, *J. Chem. Soc. Dalton Trans.* 143–147 (1972).
16. K. D. Held, H. A. Harrop, and B. D. Michael. Repair of irradiated DNA by thiols. *Int. J. Radiat. Biol.* **40**, 613–623 (1981).
17. J. R. Kanofsky. Singlet oxygen production by lactoperoxidase. Evidence from 1270 nm chemiluminescence. *J. Biol. Chem.* **258**, 5991–5993 (1983).
18. J. R. Kanofsky. Near-infrared emission in the catalase-hydrogen peroxide system, a reevaluation. *J. Am. Chem. Soc.* **106**, 4277–4278 (1984).
19. J. K. Kochi and P. J. Krusic. Isomerization and electron spin resonance of allylic radicals. *J. Am. Chem. Soc.* **90**, 7157 (1968).
20. Landolt-Börnstein. *Group II*, vol. 96, Springer-Verlag, Berlin, 1977.

21. K. R. Maples, S. J. Jordon, and R. P. Mason. In vivo rat hemoglobin thiyl free radical formation following phenylhydrazine administration. *Mol. Pharmacol.* **33,** 344–350 (1968).

22. R. P. Mason and C. Mottley. *Electron Spin Resonance,* Royal Soc. Chem. Specialist Periodical Reports, **10 B,** 185–209 (1986).

23. C. Mottley, R. P. Mason, C. F. Chignell, K. Sivarajah, and T. E. Eling. The formation of sulphur trioxide radical anion during the prostaglandin hydroperoxidase-catalyzed oxidation of bisulfite (Hydrated sulfur dioxide). *J. Biol. Chem.* **257,** 5050–5056 (1982).

24. W. G. Niehaus. A proposed role of superoxide anion as a biological nucleophile in the deesterification of phospholipids. *Bioinorg. Chem.* **7,** 77–83 (1978).

25. T. Ozawa and T. Kwan. ESR studies for chemical modification of biologically active nucleic acids by the sulphite radical anion SO_3^-. *J. Chem. Soc. Commun.* 1294–1296 (1983).

26. J. San Filippo, L. J. Romano, G. I. Chem, and J. S. Valentine. Cleavage of esters by superoxide. *J. Inorg. Chem.* **41,** 586 (1976).

27. R. D. Scurlock and P. R. Ogilby. Effect of solvent on the rate constant for the radiative deactivation of singlet molecular oxygen. *J. Phys. Chem.* **91,** 4599–4602 (1987).

28. R. C. Smith and S. J. Wyard. Electron spin resonance spectra for hydroperoxy radicals. *Nature* **186,** 226–228 (1960).

29. M. C. R. Symons. Solutions of metals: Solvated electrons. *Chem. Soc. Rev.* **5,** 337–358 (1976).

30. M. C. R. Symons. *Chemical and Biochemical Aspects of Electron Spin Resonance Spectroscopy.* Van. Nostrand Reinhold Co. Ltd., Wokingham, Berkshire, UK, 1978.

31. M. C. R. Symons. Inorganic radicals of relevance to biological systems. *Phil. Trans. R. Soc.* **B311,** 451–472 (1985).

32. M. C. R. Symons and J. M. Stephenson. Electron spin resonance studies of the solvation of superoxide ions. *J. Chem. Soc. Faraday Trans.* **177,** 1579–1583 (1981).

33. M. C. R. Symons, G. W. Eastland, L. R. Denny. Effect of solvation on the electron spin resonance of the superoxide ion. *J. Chem. Soc., Faraday, Trans.* I **76,** 1868–1874 (1980).

34. M. C. R. Symons, E. Albano, T. F. Slater, and A. Tomasi. Radiolysis of tetrachloromethane. *J. Chem. Soc., Faraday Trans.* I **78,** 2205–2214 (1982).

35. M. Trainer, M. Helten, and D. Knapska. ESR spectrum of CIO ($^2\Pi_{3/2}$) isolated in a CO_2 matrix. *J. Chem. Phys.* **79,** 3648–3655 (1983).

36. J. F. Wertz and J. R. Bolton. *Electron Spin Resonance,* McGraw Hill, New York, 1972.

From: *Trace Elements, Micronutrients, and Free Radicals* • Ed.: I. E. Dreosti • ©1991 The Humana Press Inc.

CHAPTER 2

Free Radicals as Mediators of Tissue Injury

ROLANDO DEL MAESTRO

ABSTRACT

Free radical mechanisms have been associated with a large number of disease states, such as, inflammation, ischemia-reperfusion injury, neoplasia, and aging. The elucidation of the mechanisms by which individual free radical species mediate tissue injury will further our understanding of both normal development and disease. The foci of this review are to: (1) identify the sources of free radicals in a number of disease states; (2) assess the enzymatic and nonenzymatic systems operative; and (3) review the free radical induced injury seen.

1. INTRODUCTION

Modulation of free radical reactivity is essential to the survival of aerobic organisms. This modulation involves a complex interactive process between free radical generation and a number of enzymatic and nonenzymatic systems localized to hydrophilic and hydrophobic cellular microenvironments that control these reactive species (25,35,42,56,122). In higher organisms, the controlled release of oxygen derived free radicals by specialized inflammatory cells has been harnessed to perform a bactericidal function (6,7). These same reactive species have also been implicated in the pathophysiology of a variety of disease processes, including inflammation (*25,85,90*), ischemia-reperfusion injury (*25,33,47, 65,86,87,126*), neoplasia (*46,98,117*), and aging (*39,61,62,92*).

An elucidation of the mechanisms by which individual free radical species damage biological tissue may further our understanding of normal growth and development (*1,29,30*), and a number of disease states. The focus of this review will be the role played by oxygen derived free radicals as mediators of tissue injury with an emphasis on the central nervous system.

2. DEFINITIONS

A "free radical" is defined as any atom, group of atoms, or molecules with one unpaired electron occupying an outer orbital. Molecular oxygen, O_2, is a triplet in its ground state since it has two unpaired electrons in its outer orbitals having parallel or unpaired spins ($^3\Sigma g^-$) (Table 1). Singlet oxygen ($^1\Delta g$), by definition, is not a free radical since it has both electrons occupying the same orbital and the electron spins are paired. A second excited singlet state exists ($^1\Sigma g^+$), but its short life-time suggests that it may not be an important biological species (*11*).

The parallel electron spin arrangement of O_2 prevents the direct addition of a pair of electrons (which would have electron spins in a parallel and antiparallel direction) unless an electron spin inversion occurs. The one electron reduction of O_2, called the univalent pathway, predominates, and the complete reduction of oxygen involves the addition of four electrons and four protons to each oxygen molecule (Fig. 1). The resulting intermediates are the superoxide anion radical, O_2^-, hydrogen peroxide, H_2O_2, and the hydroxyl radical, $OH^.$. Since these intermediates are very reactive, the modulation of this reactivity has been a dominating evolutionary pressure since O_2 first appeared in the atmosphere.

3. FREE RADICAL MODULATION

3.1. Scavenging Enzymes

A number of enzymatic systems have evolved that can circumvent the electron spin restriction of O_2 reduction. The cytochrome oxidase complex localized to the inner mitochondrial membrane tetravalently reduces the majority of O_2 used by aerobic cells (*3,22,25*) (Fig. 1). Mitochondria still appear to be a major intracellular source of both O_2^- and H_2O_2 (*16,17,20,22,118,119*). This continued O_2^- flux has resulted in the evolution of a variety of superoxide dismutases that function by catalytically scavenging O_2^-. In animal cells, two metalloprotein enzymes, a

Table 1
Electronic States of Oxygen

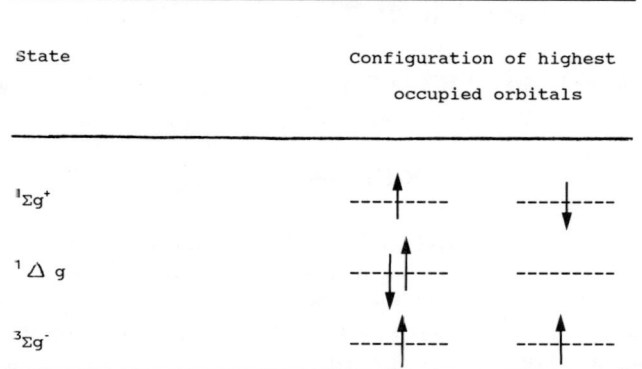

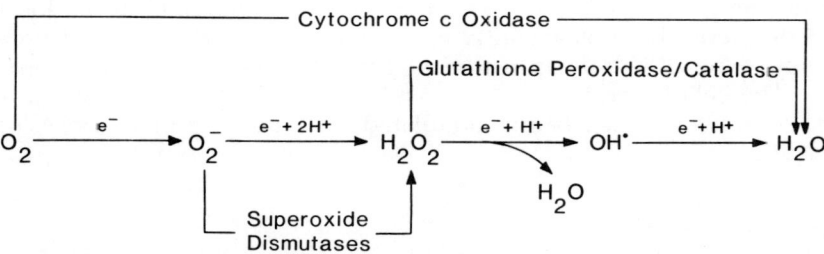

Fig. 1.   The univalent pathway for the reduction of molecular oxygen and the enzymatic defense mechanisms available to bypass and prevent the accumulation of reactive intermediates.

copper- and zinc-containing superoxide dismutase (CuZnSOD) and a manganese-containing superoxide dismutase (MnSOD) reduce O_2^- to H_2O_2 (reaction 1) (10,41–43,89). Marklund (79–81) has described and characterized a mammalian superoxide dismutase that contains copper and zinc, but whose major anatomical location appears to be the extracellular space (EC-SOD).

$$O_2^- + O_2^- + 2H^+ \longrightarrow H_2O_2 + O_2 \qquad (1)$$

Hydrogen peroxide, the divalent product of O_2 reduction, is decomposed to H_2O by catalase (CAT, reaction 2).

$$2 H_2O_2 \longrightarrow 2 H_2O + O_2 \qquad (2)$$

and a variety of peroxidases (reaction 3).

$$H_2O_2 + RH_2 \longrightarrow 2\ H_2O + R \tag{3}$$

Glutathione peroxidase (GSH-Px) has been the most intensely studied enzyme of this group (reaction 4) (38).

$$H_2O_2 + 2\ GSH \longrightarrow 2\ H_2O + GSSG \tag{4}$$

These enzymes may be envisioned as a scavenging system in which each enzyme plays an integral role in free radical modulation. A variety of scavenging enzyme patterns have been found in normal (9,29,30,78) and tumor tissues (31,82,97,98), but the reasons for this heterogeneity remains unclear. The subcellular concentration and location of these enzymes should be linked both to the site of generation of the appropriate substrate and other components of the enzyme scavenging system. In man, the genes for CuZnSOD, MnSOD, CAT, and GSH-Px are located on different chromosomes 21, 6, 11, and 3, respectively (9). Conceptually, a coordinated expression of gene activities would be expected.

The cellular and subcellular activities of scavenging enzymes in a tissue are not static, being modulated by a number of factors. These include:

1. Age;
2. Differentiation;
3. Alterations in the intracellular generation of O_2^- and H_2O_2;
4. Local availability of constituent metals; and
5. The local concentration of a number of immunomodulating agents called lymphokines.

Neonatal cerebral development in the rat is characterized by increasing CuZnSOD and MnSOD levels, a fall in CAT and, after an initial fall, increasing GSH-Px (29). Subcellular studies carried out on rat cerebral cortex demonstrated marked differences in ontogenic pattern (30). The nuclear fraction showed increasing CuZn and MnSOD activities, but low and static activities of GSH-Px and CAT during development. The mitochondrial fraction had increased MnSOD and GSH-Px activities and a rapid decrease in CAT activity during development, whereas the nonmitochondrial, nonnuclear (cytoplasmic) fraction had increasing CuZnSOD activity, decreased CAT activity, measurable but stable activities of MnSOD, and after an initial fall, increasing GSH-Px activity. A distinct ontogenetic pattern of oxidative enzyme activities and subcellular location is seen in rat cerebral cortex, and a hypothesis has been advanced to explain the patterns observed (30).

A number of investigators have hypothesized that alterations in enzymatic scavenging systems may play a role in tumor initiation and growth (46,98,117). Lowered MnSOD is seen in SV40-transformed W138 human embryonic lung fibroblasts (97) and SV40 transformed normal human skin fibroblasts (83). A murine cerebral glial tumor, C6 astrocytoma, demonstrates different cellular and subcellular contents of scavenging enzymes when compared to normal rat brain (31). Intracellular activities of scavenging enzymes increase when intracellular O_2^- and H_2O_2 rates are augmented by hyperoxia (67), paraquat (64), and possibly, hypoxia (104). A cooperative interrelationship exists between increased O_2^- flux and oxidative enzyme inactivation. The in vitro generation of increased O_2^- flux results in GSH-Px (14) and CAT (72,73) inactivation whereas CuZnSOD is inactivated by high concentrations of H_2O_2 (13). The maintenance of enzymatic scavenging capacity during states of increased O_2^- flux may be dependent on the ability of the individual enzymes to catalytically decompose their substrates to prevent inactivation of other components of the enzymatic scavenging system. Low nutritional levels of selenium appear to be reflected in low GSH-Px activity (29,30). A number of specific lymphokines, such as interleukin (α and β) (84) and tumor necrosis factor (α and β) (129), selectively increase mRNA levels for MnSOD in a variety of nonneoplastic and neoplastic cell lines, and in vivo in mice (129). The possibility exists that disease processes associated with specific lymphokine production may modulate the levels of scavenging enzyme present in a specific tissue.

In summary, although specific enzymatic scavenging systems have evolved to deal with the intracellular products of O_2 reduction, these systems should not be considered static, but capable of responding dynamically to a number of distinct intracellular levels of free radical generation.

3.2. Hydrophobic Mechanisms

The major hydrophobic regions associated with cells are their lipid membranes which contain polyunsaturated fatty acids (PUFA). Lipid peroxidation reactions can be initiated by free radical species, resulting in chain propagating lipid radical reactions which can release lipid hydroperoxides (Fig. 2) (44). A variety of hydrophobic scavengers such as tocopherols (α,β,γ,δ), collectively known as vitamin E and β-carotenes (18), intercollated into cellular membranes may prevent chain propagating reactions in lipid microenvironments. Vitamin E reacts with lipid hydroperoxides (ROO˙ and RO˙) to form a relatively stable vitamin E phenoxy radical (vit. E˙) (reaction 5) that, because of its low reactivity, does not continue chain propagation reactions (128).

$$\text{Vit E} + \text{ROO}^\bullet \rightarrow \text{ROOH} + \text{Vit E}^\bullet \qquad (5)$$

Lipid hydroperoxides are substrates for GSH-Px, but this soluble enzyme may not be able to reduce lipid hydroperoxides localized to hydrophobic

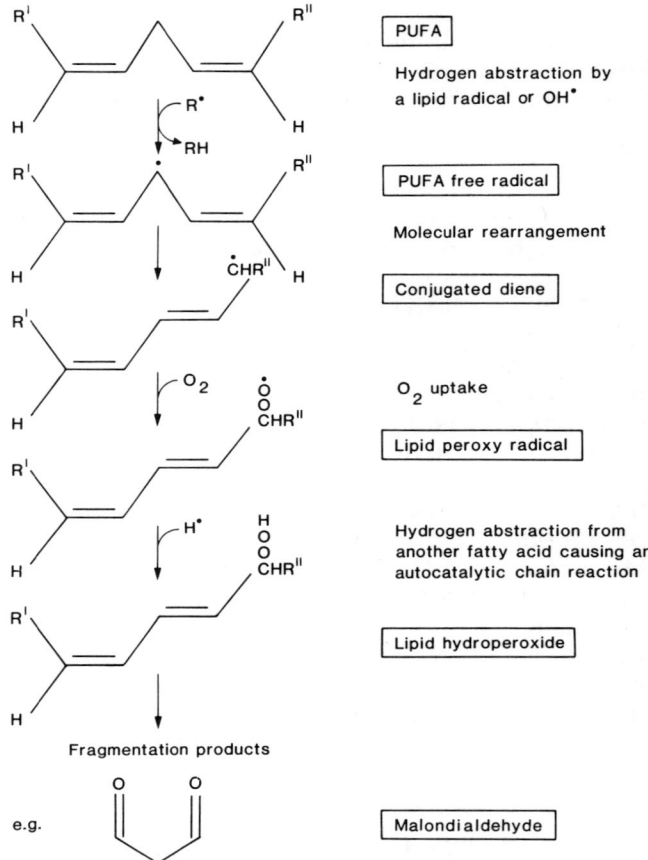

Fig. 2. Schematic representation of polyunsaturated fatty acid (PUFA) peroxidation.

microenvironments (50). The coupling of the action of specific phospholipases, which can remove lipid hydroperoxides from membrane coupled with subsequent hydroperoxide reduction by GSH-Px, may occur. A number of other GSH-dependent systems, glutathione-S-transferases (122) and a GSH-dependent protein derived from liver (120) have been suggested to play roles in hydroperoxide reduction. Both tocopherols and β-carotene may quench singlet oxygen modulating the ability of this molecule to interact with other cellular constituents.

3.3. Hydrophilic Mechanisms

Mechanisms controlling free radical reactivity in ionic or water cellular microenvironments, along with the extracellular space, have been outlined (25,122). The oxidation of extracellular, cellular, and membrane bound proteins can occur by a number of the strong oxidants generated by free radical reactions (25,122). Molecules such as ascorbic acid, cysteine, and reduced glutathione have all been considered to play a role in the prevention of this oxidation and the regeneration of normal protein structure. Ascorbic acid can regenerate the antioxidant form of vitamin E (reaction 6).

$$\text{Vit E}^{\bullet} + \text{ascorbic acid} \rightarrow \text{semidehydroascorbate radical Vit E} \quad (6)$$

However, the interaction in vivo between an antioxidant, vitamin E, which is concentrated in hydrophobic microenvironments with a hydrophilic scavenger such as ascorbic acid, is unknown (24). Ascorbic acid in the presence of iron has significant prooxidant activity (122). The role played by ascorbic acid in vivo remains to be defined. Ceruloplasmin (2) and transferrin (12) may also have scavenging capacity in vivo. Initially, it was considered that ceruloplasmin scavenged O_2^- (45). Ceruloplasmin appears to have the capacity to reduce O_2 directly to H_2O without release of reactive intermediates (cytochrome oxidase like activity) (130).

3.4. Structural Mechanisms

The reactivity of the specific free radical generated, combined with the specific biochemical microenvironment in which it is generated, determine the biomolecular and subsequent cellular injury sustained by the organism. Each intracellular organelle has a hierarchy of scavenging mechanisms localized to its individual hydrophobic and hydrophilic microregions, along with other mechanisms that may modulate free radical reactivity in that organelles' microenvironment. Intracellular scavenging systems are complex, appear interdependent on other systems, and may respond dynamically to a number of environmental alterations. Mechanisms appear to be available to allow enzymatic scavenging systems, and some "acute phase" reactants like ceruloplasmin to respond to increased intracellular endogenous or exogenous sources of free radicals. Extracellular tissues such as plasma, synovial fluid, and cerebrospinal fluid (78) possess low levels of SOD, CAT, and GSH-Px and, therefore, may be more vulnerable to free radical injury. The role played by EC-SOD in modulating O_2^- reactively extracellularly appears to be an important area for further study (79–81).

4. OXIDANT TISSUE INJURY SYSTEMS

A number of distinct oxidative tissue injury systems have been proposed to play a role in normal metabolic and disease states. Although these systems are clearly interdependent, they will be considered under a number of specific categories.

4.1. Superoxide Anion Radicals

The chemistry of O_2^- has been intensively studied (*36*). Superoxide anion radical has the potential to react as a reducing agent, donating its extra electron, or as an oxidizing agent in which it is reduced to H_2O_2. The reactivity of O_2^- with biological compounds in aqueous solution at physiological pH is limited by its spontaneous dismutation rate constant ($K \approx 2 \times 10^5 M^{-1}s^{-1}$). Although the rate of spontaneous dismutation (reaction 1) is fast, the presence of SOD increases this rate intracellularly (*42*). In microenvironments in which SOD content is low, reactions that have rate constants competitive with spontaneous dismutation would occur. A host of biological compounds, including transition metals (*57,58,88*) and quinones (*117*), can be reduced by O_2^-, and O_2^- can oxidize catecholamines (*69*), ascorbic acid (*127*), and other compounds (*36*). It has been suggested that O_2^- may be reactive in hydrophobic cellular microregions as a base, and at low pH, HO_2^-, it's protonated form, may play an oxidant role in membranes reacting with fatty acids and other hydrophobic compounds (*122*). The ability of O_2^- to cross plasma membranes via anion gaps is a mechanism by which extra-cellularly generated O_2^- may penetrate cell membranes reaching intra-cellular targets (*77*). The ability of O_2^- to diffuse relatively long distances, combined with its ability to reduce transition metals strategically located on or near important macromolecules such as DNA, may result in a selectivity of tissue macromolecular injury that has been called "site specific injury" (*21,109*).

4.1.1. Sources of O_2^-

4.1.1.1. INTRACELLULAR, MITOCHONDRIA   Mitochondria isolat-ed from a number of tissues, including brain (*22,39,100*), heart (*95,114*), lung (*118,119*), and liver (*16*), are major sources of O_2^- and H_2O_2. By spontaneous and/or enzymatic dismutation, O_2^- is an essential pre-cursor of H_2O_2 in mitochondria (*20,95*). Floyd and coworkers (*38*) have shown that malate and glutamate support O_2^- generation by intact cerebral mitochondria in vitro. A number of groups have shown that submitochondrial particles generate O_2^- (*22,119*). Succinate-supported H_2O_2 release by intact mitochondria isolated from normal adult rat brain accounted for an estimated 3% of the oxygen consumption (*22*), which is in agreement with the concept that 1–4% of oxygen reduction proceeds via the univalent pathway (*16*). When maximally reduced, both the NADH dehydrogenase and the ubiquinone-cytochrome-*b* regions of the electron transport chain are sources of H_2O_2 (Fig. 3). The influence of rotenone on succinate-supported H_2O_2 generation suggests that in the absence of inhibitors, most of the H_2O_2 generation occurs by reversed electron flow via the transfer of electrons from ubiquinone to the NADH dehydrogenase portion of the electron transport chain (*22*). Hyperoxia is

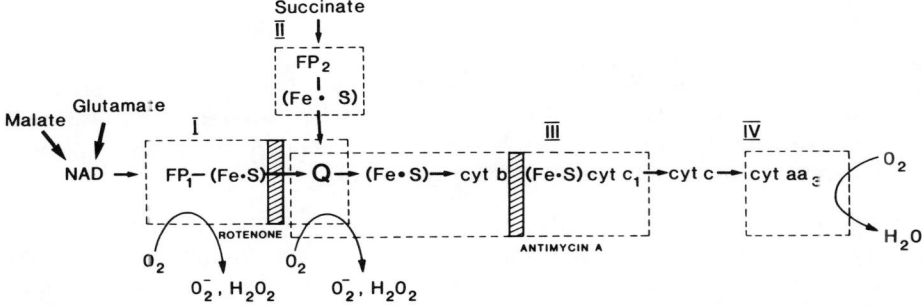

Fig. 3. The points of entry of electrons from NADH-linked substrates and succinate are shown as sites of superoxide anion radical and hydrogen peroxide generation. The sites of electron transport inhibition are outlined as hatched boxes. Abbreviations: Complex I, NADH-ubiquinone reductase; Complex II, succinate-ubiquinone reductase; Complex III, ubiquinol-cytochrome c reductase; Complex IV, cytochrome c oxidase; FP, flavoprotein; FP_1, NADH-dehydrogenase; FP_2, succinate dyhydrogenase; Fe.S, iron–sulfur center; Q, ubiquinone or coenzyme Q (reproduced with permission from 22).

associated with increased succinate-supported release of H_2O_2 by isolated cerebral (22) and lung mitochondria (118). Isolated mitochondria from a wide variety of tissues release extramitochondrial H_2O_2 as the dismutation product of at least a portion of their intramitochondrial O_2^- generation.

4.1.1.2. INTRACELLULAR, ENZYMATIC   A number of intracellular enzymes are involved in oxidation reactions in which O_2 is reduced to O_2^-. Xanthine oxidase (99) is the most extensively studied, but aldehyde oxidase, dehydroorotic dehydrogenase, flavin dehydrogenase, and peroxidases (71) are also examples. In normal tissues, xanthine oxidase exists predominately as a dehydrogenase enzyme, using NAD^+ as the electron acceptor (reaction 7).

$$\text{Hypoxanthine} + H_2O + NAD^+ \longrightarrow \text{xanthine} + NADH + H^+ \quad (7)$$

The dehydrogenase enzyme can be reversibly converted into the oxidase form by the oxidation of sulfhydryl groups, or irreversibly converted into the oxidase form by proteolysis (99). Xanthine oxidase in the presence of O_2 oxidizes hypoxanthine and xanthine to xanthine and uric acid, respectively, using O_2 as the electron acceptor (reactions 8 and 9).

$$\text{Hypoxanthine} + O_2 \longrightarrow \text{Xanthine} + O_2^- + H_2O_2 \quad (8)$$

$$\text{Xanthine} + O_2 \longrightarrow \text{uric acid} + O_2^- + H_2O_2 \quad (9)$$

At physiological pH and atmospheric O_2 concentration, about 20% of the total electron flux results in $O_2{}^-$ production. Increasing the oxygen or lowering xanthine concentrations results in increased $O_2{}^-$ and decreased H_2O_2 generation. The conversion of xanthine dehydrogenase to xanthine oxidase with the subsequent release of $O_2{}^-$ has been suggested to play a critical role in ischemic-reperfusion injury *(47,86)*.

4.1.1.3. INTRACELLULAR, AUTOXIDATION   The autoxidation of a number of intracellular compounds, including catecholamines *(69)*, flavins *(16)*, and ferredoxin *(93)* result in $O_2{}^-$ release. The contribution made by the intracellular autoxidation mechanisms to intracellular $O_2{}^-$ flux is not known.

4.1.1.4. EXTRACELLULAR, THE ACTIVATED INFLAMMATORY CELL
The discovery by Metchnikoff *(91)* of the role of the release of cytases (lysosomal products) by phagocytes in inflammation has heralded continuing research into the role of phagocytes in tissue injury. Inflammation may be defined as the organism's response to injury, and therefore, plays a role in all disease processes. Specialized inflammatory cells, such as polymorphonuclear leukocytes (PMN), macrophages, and monocytes, possess formidable bactericidal arsenals. Polymorphonuclear leukocytes release enzymes and other components from lysosomes, and specific granules release a number of oxidized arachidonic acid products (eicosinoids), and generate a flux of $O_2{}^-$ and the resultant family of reactive species *(106)*. Along with these components, macrophages, monocytes, and other immunocompetent cells may release immunomodulating agents collectively known as lymphokines *(84, 129)*, that may also play important roles in inflammation. This inflammatory arsenal can be divided into oxygen-dependent and oxygen-independent systems. Oxygen-dependent mechanisms are either dependent on O_2 derived "activated species" or on an enzyme located in primary granules, myeloperoxidase. The ability of lymphokines to induce MnSOD suggest that these compounds may modulate intracellular $O_2{}^-$ as part of their mode of action *(84,129)*.

The activation of inflammatory cells by the appropriate stimulus (bacteria, immune complexes, and so on) results in a collection of metabolic events that allow the release of $O_2{}^-$ and derived products, "the respiratory burst." Oxygen is reduced by a NADPH oxidase *(6,7)* located on the plasmalemmal membrane *(34)* of inflammatory cells. Since the NADPH oxidase is surface bound, a proportion of the $O_2{}^-$ flux is released into the extracellular space *(106)*. The ability of PMN's to reduce O_2 to $O_2{}^-$ appears essential to the killing of some types of bacteria (CAT positive) since individuals with chronic granulomatous disease in which an inherited defect in $O_2{}^-$ generation is present have repeated bacterial infection *(6,7)*.

The majority of the O_2 consumed by the respiratory burst results in $O_2{}^-$ generation *(105)* which, by spontaneous dismutation and/or SOD catalysis, accounts for the H_2O_2 generated. The mechanism(s) that result in bacterial killing by O_2^- generation is not understood, but appears to be dependent on the generation of stronger oxidants such as OH, O_2 ('Δg), and hypochlorite. Haber and Weiss (51) described a set of reactions while studying Fenton chemistry. One of these (reaction 10), the so-called Haber-Weiss reaction, was initially suggested as the mechanism by which the OH, a much stronger oxidant than O_2^- or H_2O_2 alone, could be generated by their interaction (42).

$$O_2^- + H_2O_2 \longrightarrow O_2 + OH^- + OH^\bullet \qquad (10)$$

The Haber-Weiss reaction is slow (75). However, the presence of a metal chelate involving iron or copper that can undergo reduction by O_2^- was suggested as the reaction series that occurred biologically (57,58,88), the iron catalyzed Haber-Weiss reaction (reactions 11 and 12).

$$Me^{n+} \text{ chelate} + O_2^- \longrightarrow Me^{(n-1)+} \text{ chelate} + O_2 \qquad (11)$$

$$Me^{(n-1)+} \text{ chelate} + H_2O_2 \longrightarrow Me^{n+} \text{ chelate} + OH^\bullet + OH \qquad (12)$$

It is not the focus of this review to discuss the identity of the oxidant(s) that results from a metal catalyzed interaction between O_2^- and H_2O_2. However, this oxidant or these oxidants appear to be responsible for at least a portion of the injury caused by the O_2^- and H_2O_2 interaction. The reader is referred to a number of papers that focus on this issue (58,115,127). The generation of the oxidizing species appears to be crucially dependent on the concentration and reactivity of the metal chelates present, and indeed Cu (III) and Fe (IV) have been suggested to be the oxidants involved (115).

4.1.1.5. MYELOPEROXIDASE-DEPENDENT TISSUE INJURY   Myeloperoxidase catalyzes the oxidation of a number of halides by H_2O_2 (reaction 13), and such products as hypochlorite (122,123) and O_2 ('Δg) (40,70) have been suggested as the bactericidal element of the system.

$$Cl^- + H_2O_2 \longrightarrow OCL^- + H_2O_2 \longrightarrow O_2('\Delta g) + Cl^- + H_2O \qquad (13)$$

Hydrogen peroxide generated as the result of NADPH oxidase function is used to oxidize a variety of halides (chloride, bromide, iodide) into their corresponding hypohalous acids. Hypochlorous acid is a strong oxidant and capable of oxidizing a whole series of compounds that may be present in the extracellular space or lipid and protein components of the plasmalemma. Various amino acids can be oxidatively decarboxylated by myeloperoxidase in the presence of Cl$^-$. The role played by the resultant chloramines and aldehydes has not been elucidated (123,124). Controversy continues to revolve around the issue whether or not significant amounts of O_2 ($'\Delta g$) are released during myeloperoxidase catalyzed reactions in vivo (*11,40,70*).

4.2. Hydrogen Peroxide

Although H_2O_2 may be considered a strong oxidant in high concentrations, the biological relevance of studies carried out with 1 μm–1mM concentrations is unclear. Isolated cerebral mitochondria release 200–500 pmol H_2O_2 min^{-1} mg protein^{-1} (*22*), far below the usual concentrations of H_2O_2 used in experimental studies. Hydrogen peroxide at 1mM concentration results in plasma membrane depolarization, increased plasma membrane blebbing, cell swelling, and increased plasma membrane permeability (*113*), suggesting that, although H_2O_2 has a high permeability constant across membranes, high concentrations do result in lipid membrane damage. The oxidation of some intracellular proteins such as CuZnSOD by H_2O_2 is associated with preferential degradation of this protein, which suggests that proteins altered by free radical interaction may be selectively removed from the intracellular milieu (*108*). Hydrogen peroxide, by reacting with metal chelates (reaction 12), may react in the generation of stronger oxidants, and it is an important substrate for myeloperoxidase (reaction 13).

4.2.1. Hydrogen Peroxide: Sources

Two types of reactions result in H_2O_2 generation: (1) the divalent reduction of O_2 by enzymes, such as urate-, D-amino acid, and xanthine-oxidase and (2) the spontaneous or enzyme catalyzed dismutation of O_2^-. The majority of the divalent enzymes that result in H_2O_2 generation are localized to specialized organelles called peroxisomes (*20*). The mitochondria may be one of the major intracellular sources of H_2O_2 generation, although any intracellular source of O_2^- can result in H_2O_2 production. The contribution of CAT or GSH-Px to H_2O_2 reduction has been debated, but may depend both on H_2O_2 concentration and the concentration of that particular enzyme at the site of H_2O_2 generation (*38*). The compartmentalization of CAT to peroxisomes would clearly suggest a role for this enzyme in this intracellular organelle. The presence of high concentrations of CAT in neonatal cerebral mitochondrial and cytosolic fractions (*29,30*), along with the presence of high concentrations in malignant glial cell lines and cerebral tumors (*31*), suggest that high H_2O_2 generation may play a role in undifferentiated cells (*30*). The nucleus of cerebral cells may be particularly susceptible to exogenous and endogenous sources of H_2O_2 because of low CAT and GSH-Px content.

4.3. Hydroxyl Radical

The OH$^\bullet$ is an extremely reactive and unstable oxidizing species, reacting with a wide variety of biological compounds in both hydrophobic and hydrophilic cellular microenvironments. Hydroxyl radical participate in addition, hydrogen abstraction, and electron transfer reactions (*103*). Unlike O_2^- and H_2O_2, no enzyme system has evolved to modulate OH˙ reactivity directly, and the goal of cellular scavenging mechanisms appears to be the avoidance of OH˙ generation. The potential for the generation of OH˙ or other reactive oxidants exists in any microenvironment in which O_2^- is generated, and an appropriate in vivo metal chelate exists (reactions 11 and 12).

4.4. Role of Metal Chelates

The exact nature of the metal chelates (iron and copper), which can participate in reactions 11 and 12, and their distribution in the extracellular and intracellular space is unclear. Body iron is localized to heme containing proteins, iron-binding proteins, such as transferrin, lactoferrin, ferritin, and hemosiderin, or in enzymes. Protein-bound iron may not participate in the generation of reactive oxidants by O_2^- and H_2O_2, and transferrin and lactoferrin probably also do not participate in these reactions (*see* review 12). Ferritin is the main protein involved in iron storage, and O_2^- released by PMN or xanthine oxidase has been shown to possibly enter the ferritin core through hydrophilic channels, reduce ferric iron to ferrous, and release ferrous iron such that it can participate in metal catalysis (*12*).

Iron chelated to low mol wt substances, such as ADP and citrate, have been considered to comprise a "low mol wt iron pool" in vivo. A number of studies have suggested that this low mol wt iron is increased in rheumatoid synovial fluid (*107*), although other investigators have not detected this (*12*). The role played by this low mol wt iron pool is unclear, but ferritin appears to be a reasonable candidate from which iron may be mobilized.

5. FREE RADICAL MEDIATED TISSUE INJURY

Three prerequisites appear to be necessary before confirming a role for free radicals in biomolecular injury to a specified tissue. These are: (1) that there must be a defined source of free radical species, (2) evidence that the free radical scavenging systems are either deficient or have not dynamically altered to adjust to the altered rates of radical generation, and (3) biomolecular evidence that free radical induced injury has indeed occurred.

In a previous article, disease states in which free radicals were felt to play a role were classified by the major site of pathological free radical generation (*25*). Increased free radical generation could occur extracellularly, intracellularly, or both intracellularly and extracellularly. Decreased free radical generation may occur extracellularly or intracellularly. It is impossible to review all the disease processes in which free radical tissue injury may play a role. However, by selecting disease processes that embody the general concepts, links with other disease states may be appreciated.

5.1. Increased Extracellular Free Radical Generation: Inflammatory Processes

Tissues characterized by low cell content and rich in high mol wt materials are known collectively as connective tissue. Synovial fluid is the best studied, but vascular wall, cartilage, and bone are other examples. The macromolecular components of connective tissues are made up of fibrous proteins like collagen and elastin and connective tissue polysaccharides; glycosaminoglycans like hyaluronic acid, and mucopolysaccharides, such as chondroitin sulfate. Hyaluronic acid with its high mol wt and high viscosity functions as a lubricant in joint synovial fluid. The degradation of hyaluronic acid has been a consistent feature of inflammatory arthropathies (8). The secretion products of PMN are unable to degrade hyaluronic acid, and hyaluronidases have not been found in synovial fluids (5). McCord (85) suggested that the release of free radical species from activated inflammatory cells was responsible for the hyaluronic acid degradation seen. Greenwald (48) later confirmed that PMN stimulated by phorbol myristate in the presence of ferrous iron depolymerized hyaluronic acid. Collagen structure also appears to be modified by free radical generation (49). Quantitatively, PMN's are the most important source of free radical generation in synovial fluid, and macrophage-like cells in the synovial membrane (12). In inflammatory joint disease, PMN and macrophages are activated by a variety of immune complexes in an extracellular environment containing low concentrations of SOD, CAT, and GSH-Px, to release O_2^-. Superoxide anion radical may release ferrous ion from high levels of ferritin present in these joints (107), which results in the oxidative degradation of hyaluronic acid and collagen with resultant joint dysfunction.

Another situation in which extracellular oxidation may be important in inflammatory processes is the ability of PMN to inactivate the plasma protein, a-1 proteinase inhibitor, amplifying elastase induced injury (19). Activated PMN release elastase from primary granules that can alter a number of connective tissue proteins. Plasma contains high levels of a-1 proteinase inhibitor to modulate elastase reactivity (116). Hypochlorous acid or chlorinated endogenous amines can inactivate a-1 proteinase inhibitor (19,124) that increases extracellular elastase initiated damage.

5.2. Increased Intracellular Free Radical Generation: Ischemia-Reperfusion Injury

The major difficulty in deciphering intracellular free radical mechanisms in a complex tissue such as brain or heart is the heterogeneity of the tissue. Mechanisms of measuring free radical flux from specified populations of cells making up an organ are not available. An under-

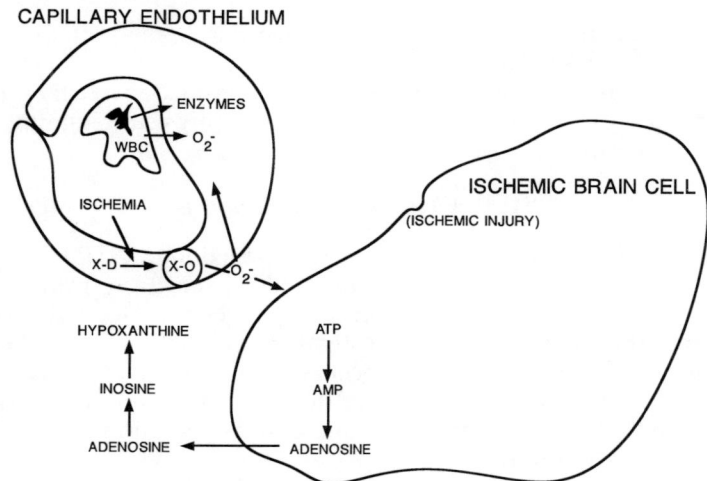

Fig. 4.   One hypothesis that has been advanced to account for reperfusion injury. Xanthine dehydrogenase (X-D) localized to capillary endothelium undergoes conversion to xanthine oxidase (X-O) induced by a Ca^{2+} stimulated protease (induced by ischemia). Xanthine oxidase acts on hypoxanthine available from purine breakdown during ischemia, and results in increased O_2^- and H_2O_2 production. Endothelial cell injury may occur as a result of increased O_2^- and/or H_2O_2 flux through the generation of more reactive intermediates, such as OH^-. Superoxide anion radical may pass through anion channels in the endothelium and result in injury to surrounding cells. Amplification of the endothelial injury may occur via the adherence and activation of inflammatory white blood cells (WBC), such as polymorphonuclear leukocytes (modified from 65).

standing of the limitations of any of the techniques used to assess the prerequisites for confirming a role for free radicals in ischemia-reperfusion injury appears essential for the interpretation of results.

Three defined free radical hypotheses have been proposed to explain a portion of the tissue injury seen in ischemia-reperfusion injury. These hypotheses are:

1. The Mitochondrial Hypothesis: During ischemia, the mitochondrial electron transport components dissociate, resulting in the accumulation of ubisemiquinone that autoxidizes on reperfusion, generating increased O_2^- and H_2O_2 that overwhelms intrinsic scavenging mechanisms and contribute to intracellular injury (33);
2. The Xanthine Dehydrogenase-Xanthine Oxidase Hypothesis: During ischemia, xanthine dehydrogenase undergoes conversion to xanthine oxidase induced by a Ca^{2+} stimulated protease (induced by ischemia). Xanthine oxidase acts on hypoxanthine available from purine breakdown, and this results in increased O_2^- and H_2O_2 generation (Fig. 4) (47,36).

3. The Inflammatory Hypothesis: During ischemia, alterations occur in tissue, which results in the accumulation and activation of inflammatory cells that amplify tissue damage by their release of reactive products (*85,87*) and physical plugging of microvessels (*111,112*).

5.2.1. Mitochondrial Hypothesis

The function of mitochondria is the production of ATP by oxidative phosphorylation, using energy generated by electron transfer from NADH to oxygen along the electron transport chain. Depending on the model and the species used, investigators have found a decrease in mitochondrial respiratory rates associated with ischemia (*22*). One of the hypotheses that have been suggested to account for mitochondrial and other intracellular tissue injury is the increased generation of free radicals from intramitochondrial sites.

Some investigators have demonstrated that ischemia is associated with an alteration of mitochondrial scavenging enzymes. Rabbit heart mitochondrial SOD decreases in ischemia (*37,114*), and similar results have been seen in human skeletal muscle (*23*). Neither complete cerebral ischemia (15 min decapitation ischemia) in the rat (*22*) or severe incomplete cerebral ischemia (8 min) followed by reperfusion are associated with significant alterations in the scavenging enzyme content in either total tissue or a mitochondrial fraction (Del Maestro, unpublished results). Further information on the role of ischemia on mitochondrial scavenging enzyme system in a number of models may aid in assessing the response of mitochondrial scavenging systems to ischemia. The hypothesis that mitochondria damaged during complete cerebral ischemia generate increased amounts of O_2^- and H_2O_2 on postischemic reoxygenation has been tested (*22,114*). Succinate-supported H_2O_2 generation by mitochondria isolated from rat brain exposed to 15 min of postdecapitative ischemia was 90% lower than control preparations (Fig. 5). Comparison of the effect of oxygen tension and respiratory chain inhibitors on succinate-supported H_2O_2 generation suggests that the ability for reversed electron transfer is impaired during ischemia. Exposure of isolated cerebral mitochondria to free radical-generating systems demonstrate that these organelles are sensitive to free radical induced injury (*66*). No evidence for lipid or protein oxidation in cerebral mitochondria isolated for brain tissue exposed to 15 min of decapitative ischemia was seen (*22*). Mitochondria isolated from rabbit heart exposed to global ischemia also had a significant decrease in succinate supported H_2O_2 release (*114*). These data do not support the hypothesis that postischemic reoxygenation is associated with increased mitochondria free radical generation. Direct assessment of this hypothesis in a number of other ischemia-reperfusion models seems warranted.

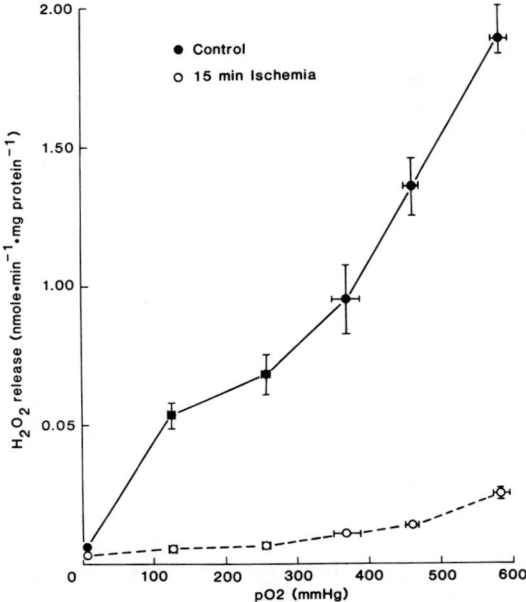

Fig. 5. Effect of oxygen on succinate-supported H_2O_2 release by mitochondria isolated from control and ischemic rat brains. Results represent mean values ±SEM (vertical bars) of six different preparations. Oxygen concentrations represent mean values ±SEM (horizontal bars) of six different determinations (reproduced with permission from 22).

5.2.2. Xanthine Dehydrogenase-Xanthine Oxidase Hypothesis

In a number of ischemia models, intestinal (47), renal (60), and cardiac (65), it has been suggested that xanthine dehydrogenase-xanthine oxidase conversion plays an important role in ischemia-reperfusion injury. The hypothesis outlined in Fig. 4 has a number of components, and combines components of free radical and Ca^{2+} induced injury (63).

5.2.2.1. SUBSTRATE   The substrates for free radical reaction must be present for their generation. Hypoxanthine and xanthine accumulate during the ischemia (110).

5.2.2.2. SITE   The distribution of xanthine dehydrogenase-xanthine oxidase among various tissues and species (99) and its distribution at a cellular level has been assessed (68). Significant variability has been found with certain organs in certain species possessing very little activity. In some human tissue such as brain and heart, these enzymes appear to be immunologically localized to capillary endothelium (*68,69*), with very little of these enzymes being present in human cerebral or cardiac cells. The presence of these enzymes located preferentially in capillary endothelium places a potential source of free radical generation at a crucial location in the microcirculation. Perturbation of microvascular flow at the capillary level may significantly influence events during reperfusion. The inhomogeneity of distribution of this enzyme system in the endothelial cells of the vascular bed and in organ systems suggests that a significant degree of variability may exist in the importance of enzymatic free radical generation in the reperfusion of specific organs in specific models after ischemia (*89*).

5.2.2.3. SYSTEM   Xanthine dehydrogenase-xanthine oxidase conversion during ischemia appears to be related to a Ca^+ activated protease (*86,99*). The ischemic induced influx of Ca^{2+} into cells may be an important trigger to prime this system for free radical generation (*63*).

5.2.2.4. INJURY AMPLIFICATION   Since purine metabolites, such as hypoxanthine, may accumulate in cells in which xanthine dehydrogenase-xanthine oxidase activities may be low, these substrates leave these cells, and a concentration gradient may exist in which these substrates enter endothelial cells where they are oxidized during reperfusion, and result in endothelial membrane alterations, resulting in permeability changes (*27,28,59,74,76,125*) and leukocyte adhesion (*26,32*).

The xanthine dehydrogenase-xanthine oxidase hypothesis has certainly increased our understanding of the role played by the enzymatic generation of free radical species during ischemic-reperfusion injury in a number of models. However, the role played by this mechanism in human ischemic states remains to be defined.

5.2.3. Inflammatory Hypothesis

Inflammation and ischemia may be linked in a synergistic manner through the secondary accumulation and activation of inflammatory cells, as partially outlined in Fig. 4. The attraction of inflammatory cells to the microcirculation of tissues involved in ischemia-reperfusion appears to be a common phenomenon (*87,126*). The depletion of PMN from animal models of ischemia appears to ameliorate injury (*126*). Inflammatory cells recognize receptors present on specific endothelial cells (*126*), and are also attracted by a number of chemoattractants. The release of O_2^- results in the formation of an O_2^- dependent chemotaxic factor from a latent complex present in the extracellular space (*101*), and the in vitro incubation of purified arachidonic acid with an O_2^- enzymatic generating system resulted in the formation of chemotactic products (*102*). In the hamster cheek pouch model, the attraction and adhesion of granulocytes was dependent on an O_2^- derived product (*26,32*). The adhesion and plugging (*111,112*) of microvascular bed by PMN may have important consequences to flow and permeability in these networks. There appears to be an important component of ischemic-reperfusion injury that is related to the role played by activated inflammatory cells in the microvasculature.

5.2.4. Influence of an Inhibitor
of Free Radical Induced Lipid Peroxidation

A number of investigations have demonstrated that an inhibitor of free radical induced lipid peroxidation, 21- 4-(2,b-di-1-pyrrolidinyl-4-pyrimidinyl-1-piperazinyl)-16a-methylpregna-1,4,9(11)-triene-e,20-dione monomethane sulfonate (U74006F)(*15*) modulates both ischemic (*52–54*) and traumatic injuries (*55*) that occur in the central nervous system in animal models. Blood brain barrier injury following experimental subarachnoid hemorrhage in the rat has been decreased by this compound (*121,131*). Although the efficacy of an inhibitor of lipid peroxidation to modulate ischemic injury in experimental models does not prove a role for free radical processes, it certainly may provide further insight into free radical induced injury in tissues.

5.3. Aging

Aging is a progressive accumulation of changes with time that, in human, occurs over decades. Harman (*61,62*) postulated that free radicals play a key role in the aging cascade. This concept has been modified to suggest that progressive free radical induced mitochondrial injury in postmitotic cells is central to the aging process (*92*).

The capacity of isolated rat heart mitochondria to release both O_2^- and H_2O_2 increases with age (*94*). Isolated rat cerebral mitochondria appear to not only increase O_2^- and H_2O_2 generation with age, but alter the proportion of these two reactive metabolites (Del Maestro and McDonald, unpublished results). Isolated heart mitochondria subjected to hyperbaric oxygenation demonstrated hydroperoxide accumulation and decreased cardiolipin and other fatty acid components of their membrane (*96*). The relationship of these high O_2^- and H_2O_2 flux rates to the much more moderate alterations seen with aging remains to be determined. The scavenging enzyme content of cerebral tissue with age has been assessed (*29*). No change in rat or mouse brain total SOD activity was found between 8 and 32 mo of age. Patterns of subcellular distribution of scavenging enzyme associated with aging have not been defined for most tissues, but appear pertinent to the aging of mitochondria.

Two autofluorescent lipopigments, lipofuscin, an inert complex that accumulates with age normally in all tissues, and ceroid, that is prominent in a number of disease processes (*4*), have been described. These lipopigments are felt to originate from the free radical attack on PUFA, resulting in the metal catalyzed decomposition of lipid hydroperoxides to yield carbonyls, which then crosslink with primary amines and undergo polymerization (4). The free radical theory of aging has provided valuable insights into the aging process and continued testing of this hypothesis may lead to further understanding of this universal phenomenon.

6. SUMMARY

Although this communication has been able to cover only some mechanisms associated with free radical mediated tissue injury, a number of concepts have been addressed. The author believes that it is essential that, before ascribing a role for free radicals in any tissue injury system, care must be taken to:

1. Identify and quantitate sources of free radical production;
2. Measure the enzymatic and nonenzymatic scavenging system operative; and
3. Quantitate free radical induced injury to proteins and lipids.

If these three prerequisites are carried out in the model system being studied, hypotheses can be adequately tested, and the field of free radical research will be continually advanced.

ACKNOWLEDGMENT

The author wishes to thank the collaborators in these studies, Maria Cino, Christopher Viereck, Janet MacKinnon, Warren McDonald, and Eric Stroude, and Jo-Ann Dunn for her excellent secretarial assistance.

These studies were supported by the Heart and Stroke Foundation of Ontario and the Brain Research Fund Foundation.

Dr. Del Maestro is a recipient of an Ontario Ministry of Health Career Scientist Award.

REFERENCES

1. R. G. Allen, and A. K. Balin. Oxidative influence on development and differentiation: An overview of a free radical theory of development. *J. Free Rad. Biol. Med.* **6,** 631–661 (1989).
2. D. J. Al-Timimi and T. L. Dormandy. The inhibition of lipid autoxidation by human caeruloplasmin. *Biochem. J.* **168,** 283–288 (1977).
3. E. Antonini, M. Brunori, C. Greenwood and B. G. Malmstrom. Catalytic mechanism of cytochrome oxidase. *Nature* **228,** 936, 937 (1970).
4. D. Armstrong. Free radical involvement in the formation of lipopigments. In *Free Radicals in Molecular Biology, Aging and Disease*, D. Armstrong, R. S. Sohal, R. G. Cutler, and T. F. Slater, eds., Raven, New York, 1984, pp. 129–141.
5. N. N. Aronson Jr. and E. A. Davidson. Lysosomal hyaluronidase. *J. Biol. Chem.* **240,** PC3222–3224 (1965).

6. B. M. Babior. Oxygen-dependent microbial killing by phagocytes. Part 1. *N. Eng. J. Med.* **298,** 659–668 (1978).
7. B. M. Babior. Oxygen-dependent microbial killing by phagocytes. Part 2. *N. Eng. J. Med.* **298,** 721–725 (1978b).
8. E. A. Balazs. The physical properties of synovial fluid and the special role of hyaluronic acid. In *Disorders of the Knee,* J. Helfet, ed., J. P. Lippincott Co., Philadelphia, 1974, pp. 63–75.
9. W. H. Bannister and J. V. Bannister. Factor analysis of the activities of superoxide dismutase, catalase and glutathione peroxidase in normal tissues and neoplastic cell lines. *Free Rad. Res. Commun.* **4,** 1–13 (1987).
10. C. O. Beauchamp and I. Fridovich. Isozymes of superoxide dismutase from wheat germ. *Biochim. Biophys. Acta* **317,** 50–64 (1973).
11. D. Belluš. Quenchers of singlet oxygen—A critical review. In *Singlet Oxygen. Reactions With Organic Compounds and Polymers,* B. Ranby and J. F. Rabek, eds., John Wiley, New York, 1978, pp. 86, 87.
12. P. Biemond, A. J. G. Swaak, H. G. Van Eijk and J. F. Koster. Superoxide dependent iron release from ferritin in inflammatory diseases. *J. Free Radical Biol. Med.* **4,** 185–198 (1988).
13. D. M. Blech and C. L. Borders. Hydroperoxide anion HO_2^- is an affinity reagent for the inactivation of yeast Cu, Zn superoxide dismutase: Modification of one histidine per subunit. *Arch. Biochem. Biophys.* **224,** 579–586 (1983).
14. J. Blum and I. Fridovich. Inactivation of glutathione peroxidase by superoxide radical. *Arch. Biochem. Biophys.* **240,** 500–508 (1985).
15. J. M. Braughler, J. F. Pregenzer, R. L. Chase, L. A. Duncan, E. J. Jacobsen, and J. M. McBall. Novel 21-amino steroids as potent inhibitors of iron dependent lipid peroxidation. *J. Biol. Chem.* **262,** 10, 438–10, 440 (1987).
16. A. Boveris. Mitochondrial production of superoxide radical and hydrogen peroxide. *Adv. Exp. Med. Biol.* **78,** 67–82 (1977).
17. A. Boveris, N. Oshino, and B. Chance. The cellular production of hydrogen peroxide. *Biochem. J.* **128,** 617–630 (1972).
18. G. W. Burton and K. U. Ingold. β-carotene: An unusual type of lipid antioxidant. *Science* **224,** 569–573 (1984).
19. H. Carp and A. Janoff. Phagocyte-derived oxidant suppress the elastase-inhibitory capacity of alpha-1-proteinase inhibitor in vitro. *J. Clin. Invest.* **66,** 987–995 (1980).
20. B. Chance, H. Sies and A. Boveris. Hydroperoxide metabolism in mammalian organs. *Physiol. Rev.* **59,** 527–605 (1979).
21. M. Chevion. A site-specific mechanism for free radical induced biological damage: The essential role of redox-active transition metals. *Free Rad. Biol. Med.* **5,** 27–37 (1989).
22. M. Cino and R. F. Del Maestro. Generation of hydrogen peroxide by brain mitochondria: The effects of reoxygenation following postdecapitative ischemia. *Arch. Biochem. Biophys.* **269,** 623–638 (1989).
23. G. G. Corbucci, A. Gasparetta, A. Candiani, G. Crimi, M. Antonelli, M. Bufi, R. A. De Blasi, M. B. Cooper, and K. Gohil. Shock-induced damage to mitochondrial function and cellular antioxidant mechanisms in humans. *Circ. Shock* **15,** 15–26 (1985).
24. R. T. Dean, and K. H. Cheeseman. Vitamin E protects against free radical damage in lipid environments. *Biochem. Biophys. Res. Commun.* **148,** 1277–1282 (1987).
25. R. F. Del Maestro. An approach to free radicals in medicine and biology. *Acta Physiol. Scand. Suppl.* **492,** 153–168 (1980).

26. R. F. Del Maestro. Role of superoxide anion radicals in microvascular permeability and leukocyte behavior. *Can. J. Physiol. Pharmacol.* **60**, 1406–1414 (1982).
27. R. F. Del Maestro, J. Bjork, and K.-E. Arfors. Increase in microvascular permeability induced by enzymatically generated free radicals. I. In vivo study. *Microvasc. Res.* **22**, 239–254 (1981a).
28. R. F. Del Maestro, J. Bjork, and K.-E. Arfors. Increase in microvascular permeability induced by enzymatically generated free radicals. II. Role of superoxide anion radical, hydrogen peroxide, and hydroxyl radicals. *Microvasc. Res.* **22**, 255–270 (1981b).
29. R. F. Del Maestro, and W. McDonald. Distribution of superoxide dismutase, glutathione peroxidase, and catalase in developing rat brain. *Mech. Aging Dev.* **41**, 29–38 (1987).
30. R. F. Del Maestro and W. McDonald. Subcellular localization of superoxides, glutathione peroxidase, and catalase in developing rat cerebral cortex. *Mech. Aging Dev.* **48**, 15–31 (1989).
31. R. F. Del Maestro, W. McDonald, and R. Anderson. Superoxide dismutases, catalase, and glutathione peroxidase in experimental and human brain tumours. In *Oxy Radicals and Their Scavenging Systems* (vol. 2), R. Greenwald and G. Cohen, eds., Elsevier, Amsterdam, 1983, pp. 28–35.
32. R. F. Del Maestro, M. Planker, and K. -E. Arfors. Evidence for the participation of superoxide anion radical in altering the adhesive interaction between granulocytes and endothelium in vivo. *Int. J. Microcirc.: Clin. Exp.* **1**, 105–120 (1982).
33. H. B. Dempoulos, E. S. Flamm, D. D. Pietronigro, and M. L. Seligman. The free radical pathology and the microcirculation in the central nervous system. *Acta Scand. Suppl.* **492**, 91–120 (1980).
34. B. Dewald, M. Baggiolini, J. T. Curnutte, and B. M. Babior. Subcellular localization of the superoxide forming enzyme in human neutrophils. *J. Clin. Invest.* **63**, 21–29 (1979).
35. T. L. Dormandy. Free radical oxidations and antioxidants. *Lancet* **II**, 647–650 (1978).
36. J. A. Fee and J. S. Valentine. Chemical and physical properties of superoxide. In *Superoxide and Superoxide Dismutases*, A. M. Michelson, J. M. McCord, and I. Fridovich, eds., Academic, New York, 1977, pp. 19–60.
37. R. Ferrari, C. Ceconi, S. Curello, C. Guarnieri, C. M. Caldarera, A. Albertini, and O. Visioli. Oxygen-mediated myocardial damage during ischemia and reperfusion: role of the cellular defenses against oxygen toxicity. *J. Mol. Cell Cardiol.* **17**, 937–945 (1985).
38. L. Flohé. Glutathione peroxidase brought into focus. In *Free Rad. in Biol.*, vol. 5, W. Pryor, ed., Academic, New York, 1982, pp. 295–319.
39. R. A. Floyd, M. M. Zaleska, and H. J. Harmon. Possible involvement of iron and oxygen free radicals in aspects of aging in brain. In *Free Radicals in Molecular Biology, Aging and Disease*, D. Armstrong, ed., Raven Press, New York, 1984, pp. 143–161.
40. C. S. Foote, R. B. Abakerli, R. L. Clough, and F. C. Shook. On the question of singlet oxygen production in leukocytes, macrophages and the dismutation of superoxide anion. In *Biological and Clinical Aspects of Superoxide and Superoxide Dismutase*, vol. 11B, W. H. Bannister and J. V. Bannister, eds., Elsevier/North Holland, New York, 1980, pp. 222–230.
41. I. Fridovich. Superoxide dismutases. *Ann. Rev. Biochem.* **44**, 147–159 (1975).
42. I. Fridovich. The biology of oxygen radicals. *Science* **201**, 875–880 (1978).

43. I. Fridovich. Superoxide radical: An endogenous toxicant. *Ann. Rev Pharmacol. Toxicol.* **23,** 239–257 (1983).
44. H. W. Gardner. Oxygen radical chemistry of polyunsaturated fatty acids. *Free Rad. Biol. Med.* **7,** 65–86 (1983).
45. I. M. Goldstein, H. B. Kaplan, H. S. Edelson, and G. Weissman. Ceruloplasmin: A scavenger of superoxide anion radicals. *J. Biol. Chem.* **254,** 4040–4045 (1979).
46. J. D. Gower. A role for dietary lipids and antioxidants in the activation of carcinogens. *Free Rad. Biol. Med.* **5,** 95–111 (1988).
47. D. N. Granger, G. Rutili, and J. M. McCord. Superoxide radicals in feline intestinal ischemia. *Gastroenterology* **78,** 474–480 (1981).
48. R. A. Greenwald. Effects of oxygen-derived free radicals on connective tissue macromolecules. In *Biological and Clinical Aspects of Superoxide and Superoxide Dismutase,* vol. 11B, W. H. Bannister and J. V. Bannister, eds., Elsevier/North Holland, New York, 1980, pp. 160–171.
49. R. A. Greenwald, and W. W. Moy. Inhibition of collagen gelatin by action of the superoxide radical. *Arthritis Rheum.* **22,** 251–259 (1979).
50. A. Grossman and A. Wendel. Nonreactivity of the selenoenzyme glutathione peroxidase with enzyme-generated hydroperoxide phospholipids. *Eur. J. Biochem.* **135,** 549–552 (1984).
51. F. Haber and J. Weiss. The catalytic decomposition of hydrogen peroxide by iron salts. *Proc. Roy. Soc. Ser. A* **147,** 332–351 (1934).
52. E. D. Hall. Effects of the 21-aminosteroid U74006F on posttraumatic spinal cord ischemia. *J. Neurosurg.* **68,** 462–465 (1988).
53. E. D. Hall, K. P. Berry, and J. M. Braughler. The 21-aminosteroid lipid peroxidation inhibitor U74006F protects against cerebral ischemia in gerbils. *Stroke* **19,** 997–1002 (1988).
54. E. D. Hall, and D. A. Yonkers. Attenuation of postischemic cerebral hypoperfusion by the 21-aminosteroid U74006F. *Stroke* **19,** 340–344 (1988).
55. E. D. Hall, D. A. Yonkers, J. M. McCall, and J. M. Braughler. Effects of the 21-aminosteroid U74006F on experimental head injury in mice. *J. Neurosurg.* **68,** 456–461 (1988).
56. B. Halliwell. Biochemical mechanisms accounting for the toxic action of oxygen on living organisms: The key role of superoxide dismutase. *Cell Biol. Int. Rep.* **2,** 113–128 (1978a).
57. B. Halliwell. Superoxide-dependent formation of hydroxyl radicals in the presence of iron salts. *FEBS Lett.* **96,** 238–242 (1978b).
58. B. Halliwell. Superoxide-dependent formation of hydroxyl radicals in the presence of iron chelates. Is it a mechanism for hydroxyl radical production in biological systems? *FEBS Lett.* **92,** 321–326 (1978c).
59. B. Halliwell. Invited Commentary. Superoxide, iron, vascular endothelium, and reperfusion injury. *Free Rad. Res. Commun.* **5,** 315–318 (1989).
60. R. Hansson, B. Gustafsson, O. Jonsson, S. Lundstam, S. Pettersson, T. Schersten, and J. Waldenstrom. Effect of xanthine oxidase inhibition on renal circulation after ischemia. *Transplantation Proc.* **14,** 51–58 (1982).
61. D. Harman. Aging: A theory based on free radical and radiation chemistry. *Univ. Cal. Rad. Lab. Rep.* No. 3078 (1955).
62. D. Harman. Aging: A theory based on free radical and radiation chemistry. *J. Gerontol.* **11,** 298–300 (1956).
63. W. K. Hass. Beyond cerebral blood flow, metabolism, and ischemic thresholds: Examination of the role of calcium in the initiation of cerebral infarction. In *Cerebral Vascular Disease,* vol. 3, Proceedings of the 10th Salzburg

Conference on Cerebral Vascular Disease, J. S. Meyer, H. Lechner, M. Reivich, E. O. Ott, and A. Arabinar, eds., Excerpta Medica, Amsterdam, 1981, pp. 3–17.

64. H. M. Hassan. Biosynthesis and regulation of superoxide dismutases. *Free Rad. Biol. Med.* **5**, 377–385.

65. D. J. Hearse, A. S. Manning, J. M. Downey, and D. M. Yellon. Xanthine oxidase: A critical mediator of myocardial injury during ischemia and reperfusion? *Acta Physiol. Scand. Suppl.* **548**, 65–78 (1986).

66. L. Hillered and L. Ernster. Respiratory activity of isolated rat brain mito-chondria following in vitro exposure to oxygen radicals. *J. Cereb. Blood Flow Metabol.* **3**, 207–214 (1983).

67. D. Jamieson, B. Chance, E. Cadenas, and A. Boveris. The relation of free radical production to hyperoxia. *Ann. Rev. Physiol.* **48**, 703–719 (1986).

68. E. D. Jarasch, G. Bruder, and H. W. Heid. Significance of xanthine oxidase in capillary endothelial cells. *Acta Physiol. Scand. Suppl.* **548**, 39–46 (1986).

69. S. L. Jewett, L. J. Eddy, and P. Hochstein. Is the autoxidation of cate-cholamines involved in ischemia-reperfusion injury? *Free Rad. Biol. Med.* **6**, 185–188.

70. J. R. Kanofsky, J. Wright, G. E. Miles-Richardson, and A. I. Tauber. Biochemical requirements for singlet oxygen production by purified human myeloperoxidase. *J. Clin. Invest.* **74**, 1489–1495 (1984).

71. H. Kohler, and H. Jenzer. Interaction of lactoperoxidase with hydrogen peroxide. Formation of enzyme intermediates and generation of free radi-cals. *J. Free Rad. Biol. Med.* **6**, 323–339 (1989).

72. Y. Kono, and I. Fridovich. Superoxide radical inhibits catalase. *J. Biol. Chem.* **257**, 5751–5754 (1982).

73. Y. Kono, and I. Fridovich. Isolation and characterization of the pseu-docatalase of lactobacillus plantarum. *J. Biol. Chem.* **258**, 6015–6019 (1983).

74. H. A. Kontos. Oxygen radicals in cerebral vascular injury. *Circ. Res.* **57**, 508–516 (1985).

75. W. H. Koppenol, J. Butler, and J. W. van Leeuwen. The Haber-Weiss cycle. *Photochem. Photobiol.* **28**, 655–660 (1978).

76. K. Ley and K.-E. Arfors. Changes in macromolecular permeability by intravascular generation of oxygen derived free radicals. *Microvas. Res.* **24**, 25–33 (1982).

77. R. E. Lynch, and I. Fridovich. Effects of superoxide on the erythrocyte membrane. *J. Biol. Chem.* **253**, 1838–1845 (1978).

78. S. L. Marklund. Distribution of CuZn superoxide dismutase and Mn super-oxide dismutase in human tissues and extracellular fluids. *Acta Physiol. Scand. Suppl.* **492**, 19–23 (1980).

79. S. L. Marklund. Human copper-containing superoxide dismutase of high mol wt. *Proc. Natl. Acad. Sci. USA* **79**, 7634–7638 (1982).

80. S. L. Marklund. Extracellular superoxide dismutase and other superoxide dismutase isoenzymes in tissues from nine mammalian species. *Biochem. J.* **222**, 649–655 (1984).

81. S. L. Marklund. Extracellular superoxide dismutase in human tissues and human cell lines. *J. Clin. Invest.* **74**, 1398–1403 (1984).

82. S. L. Marklund, N. G. Westman, E. Lundgren, and G. Roos. Copper- and zinc-containing superoxide dismutase, manganese-containing superoxide dismutase, catalase, and glutathione peroxidase in normal and neoplastic human cell lines and normal human tissues. *Cancer Res.* **42**, 1955–1961 (1982).

83. F. Marlhens, A. Nicole, and P. M. Sinet. Lowered levels of translatable messenger RNAs for manganese superoxide dismutase in human fibroblasts transformed by SV40. *Biochem. Biophys. Res. Commun.* **129**, 300–305 (1985).
84. A. Masuda, D. L. Long, Y. Kobayashi, E. Appella, J. J. Oppenheim, and K. Matsushima. Induction of mitochondrial manganese superoxide dismutase by interleukin 1. *FASEB J.* **2**, 3087–3091 (1988).
85. J. M. McCord. Free radicals and inflammation: Protection of synovial fluid by superoxide dismutase. *Science* **185**, 529–531 (1974).
86. J. M. McCord. Oxygen-derived free radicals in postischemic tissue injury. *N. Engl. J. Med.* **312**, 159–163 (1985).
87. J. M. McCord. Free radicals and myocardial ischemia: Overview and outlook. *J. Free Rad. Biol. Med.* **4**, 9–14 (1988).
88. J. M. McCord and E. D. Day. Superoxide-dependent production of hydroxyl radical catalyzed by iron-EDTA complex. *FEBS Lett.* **86**, 139–142 (1978).
89. J. M. McCord and I. Fridovich. Superoxide dismutase: An enzymatic function for erythrocuprein (hemocuprein). *J. Biol. Chem.* **244**, 6049–6055 (1969).
90. J. M. McCord, K. Wong, S. H. Stokes, W. F. Petrone, and D. English. Superoxide and inflammation: A mechanism for the antiinflammatory activity of superoxide dismutase. *Acta Physiol. Scand. Suppl.* **492**, 25–30 (1980).
91. E. Metchnikoff. Immunity in infective diseases. Johnson Reprint Corp., New York and London, 1905.
92. J. Miquel and J. E. Fleming. A two-step hypothesis on the mechanism of in vitro cell aging: Cell differentiation followed by intrinsic mitochondrial mutagenesis. *Exp. Gerontol.* **19**, 31–36 (1984).
93. H. P. Misra, and I. Fridovich. The generation of superoxide radical during the autoxidation of ferredoxin. *J. Biol. Chem.* **240**, 6886–6890 (1971).
94. H. Nohl. Oxygen radical release in mitochondria: Influence of age. In *Free Radicals, Aging and Degenerative Diseases*, J. E. Johnson, Jr., R. Walford, D. Harman, and J. Miguel, eds., Alan R. Liss, New York, 1986, pp. 77–97.
95. H. Nohl and D. Hegner. Do mitochondria produce oxygen radicals in vitro? *Eur. J. Biochem.* **82**, 563–567 (1978).
96. H. Nohl, D. Hegner, and K. H. Summer. The mechanism of toxic action of hyperbaric oxygenation on the mitochondria of rat heart cells. *Biochem. Pharmacol.* **30**, 1753–1757 (1981).
97. L. W. Oberley, M. L. McCormick, E. Sierra-Rivera, and D. Kasemset-St. Clair. Manganese superoxide dismutase in normal and transformed human embryonic lung fibroblasts. *J. Free Rad. Biol. Med.* **6**, 379–384.
98. L. W. Oberley, T. D. Oberley, and G. R. Buettner. Cell division in normal and transformed cells: The possible role of superoxide and hydrogen peroxide. *Med. Hypotheses* **7**, 21–42 (1981).
99. D. A. Parks, and D. N. Granger. Xanthine oxidase: Biochemistry, distribution and physiology. *Acta Physiol. Scand. Suppl.* **548**, 87–99 (1986).
100. M. S. Patole, A. Swaroop, and T. Ramasarma. Generation of H_2O_2 in brain mitochondria. *J. Neurochem.* **47**, 1–8 (1986).
101. H. D. Perez, and I. M. Goldstein. Generation of a chemotactic lipid from arachidonic acid by exposure to a superoxide-generating system. *Fed. Proc.* **38**, 1170 (1979).
102. H. D. Perez, B. B. Weksler, and I. M. Goldstein. Generation of a chemotactic lipid from arachidonic acid by exposure to a superoxide-generating system. *Inflammation* **4**, 313–328 (1980).

103. W. A. Pryor. The role of free radical reactions in biological systems. In *Free Radicals in Biology*, vol. 1, W. Pryor, ed., Academic, New York, 1976, pp. 1–49.

104. M. Rister, and R. L. Baehner. The alteration of superoxide dismutase, catalase, glutathione peroxidase and NAD(P)H cytochrome-c reductase in guinea pig polymorphonuclear leukocytes and alveolar macrophages during hypoxia. *J. Clin. Invest.* **58**, 1174–1184 (1976).

105. R. K. Root and J. A. Metcalf. H_2O_2 release from human granulocytes during phagocytosis. Relationship to superoxide anion formation and cellular catabolism of H_2O_2: Studies with normal and cytochalasin B-treated cells. *J. Clin. Invest.* **60**, 1266–1279 (1977).

106. H. Rosen, and S. J. Klebanoff. Bactericidal activity of a superoxide anion generating system. A model for the polymorphonuclear leukocyte. *J. Exp. Med.* **149**, 27–39 (1979).

107. D. Rowley, J. M. C. Gutteridge, D. Blake, M. Farr, and B. Halliwell. Lipid peroxidation in rheumatoid arthritis: Thiobarbituric acid reactive material and catalytic iron salts in synovial fluid from rheumatoid arthritis patients. *Clin. Sci.* **66**, 691–695 (1984).

108. D. C. Salo, S. W. Lin, R. E. Pacifici, and K. J. A. Davies. Superoxide dismutase is preferentially degraded by a proteolytic system from red blood cells following oxidative modification by hydrogen peroxide. *J. Free Rad. Biol. Med.* **5**, 335–339 (1988).

109. A. Sammuni, M. Chevion, and G. Czapski. Unusual copper-induced sensitization of the biological damage due to superoxide radicals. *J. Biol. Chem.* **256**, 12632–12635 (1981).

110. O. D. Saugstad, and A. O. Aasen. Plasma hypoxanthine concentrations in pigs—A prognostic aid in hypoxia. *Eur. Surg. Res.* **12**, 123–129 (1980).

111. G. W. Schmid-Schonbein. Capillary plugging by granulocytes and the no-reflow phenomenon in the microcirculation. *Fed. Proc.* **46**, 2397–2401.

112. G. W. Schmid-Schoenbien, and R. L. Engler. Leukocyte capillary plugging in myocardial ischemia and during reperfusion in the dog. *Am. J. Pathol.*: 98–111 (1983).

113. J. A. Scott, A. J. Fischman, J. Homey, J. T. Fallon, B. A. Khan, C. A. Peto, and C. A. Rabito. Morphologic and functional correlates of plasma membrane injury during oxidant exposure. *J. Free Rad. Biol. Med.* **6**, 361–367 (1989).

114. M. Shlafer, C. L. Myers, and S. Adkins. Mitochondrial hydrogen peroxide generation and activities of glutathione peroxidase and superoxide dismutase following global ischemia. *J. Mol. Cell Cardiol.* **19**, 1195–1206 (1987).

115. H. C. Sutton, and C. C. Winterbourn. On the participation of higher oxidation states of iron and copper in Fenton reactions. *J. Free Rad. Biol. Med.* **6**, 53–60 (1989).

116. J. Travis, and G. S. Salvesen. Human plasma proteinase inhibitors. *Ann. Rev. Biochem.* **52**, 655–709 (1983).

117. P. O. P. Ts'o, W. J. Caspary, and R. J. Lorentzen. The involvement of free radicals in chemical carcinogenesis. In *Free Radicals in Biology*, vol. 3, W. Pryor, ed., Academic, New York, 1977, pp. 251–303.

118. J. F. Turrens, B. A. Freeman, and J. D. Crapo. Hyperoxia increases H_2O_2 release by lung mitochondria and microsomes. *Arch. Biochem. Biophys.* **217**, 411–421 (1982)

119. J. F. Turrens, B. A. Freeman, J. G. Levitt, and J. D. Crapo. The effect of hyperoxia on superoxide production by lung submitochondrial particles. *Arch. Biochem. Biophys.* **217**, 401–410 (1982).

120. F. Ursini, M. Maiorino, M. Valente, L. Ferri, and C. Gregolin. Purification from pig liver of a protein which protects liposomes and biomembranes from peroxidative degradation and exhibits glutathione peroxidase activity on phosphatidylcholine hydroperoxides. *Biochem. Biophys. Acta* **710,** 197–211 (1982).
121. D. G. Vollmer, N. F. Kassell, K. Hongo, H. Ogawa, and T. Tsukahara. Effect of the nonglcocorticoid 21-aminosteroid U74006F on experimental cerebral vasospasm. *Surg. Neurol.* **31,** 190–194 (1989).
122. S. J. Weiss. Oxygen, ischemia and inflammation. *Acta Physiol. Scand. Suppl.* **548,** 9–37 (1986).
123. S. J. Weiss, R. Klein, A. Slivka, and M. Wei. Chlorination of taurine by human neutrophils: Evidence for hypochlorous acid generation. *J. Clin. Invest.* **70,** 598–607 (1982).
124. S. J. Weiss, M. B. Lampert, S. T. Test. Long-lived oxidants generated by human neutrophils characterization and bioactivity. *Science* **222,** 625–628 (1983).
125. S. J. Weiss, J. Young, A. F. LoBuglio, A. Slivka, and N. F. Nimek. Role of hydrogen peroxide in neutrophil-mediated destruction of cultured endothelial cells. *J. Clin. Invest.* **68,** 714–721 (1981).
126. S. W. Werns, and B. R. Lucchesi. Leukocytes, oxygen radicals and myocardial injury due to ischemia and reperfusion. *Free Rad. Biol. Med.* **4,** 31–37 (1987).
127. C. C. Winterbourn. Hydroxyl radical production in body fluids. Roles of metal ions, ascorbate and superoxide. *Biochem. J.* **198,** 125–131 (1981).
128. L. A. Witting. Vitamin E and lipid antioxidants in free radical initiated reactions. In *Free Radicals in Biology,* W. Pryor, ed., Academic, New York, 1982, pp. 295–319.
129. G. H. W. Wong, and D. V. Goeddel. Induction of manganese superoxide dismutase by tumour necrosis factor: Possible protective mechanism. *Science* **242,** 941–944 (1988).
130. S. Yamashogi and G. Kajimoto. Antioxidant effect of ceruloplasmin on microsomal lipid peroxidation. *FEBS Lett.* **152,** 168 (1983).
131. M. Zuccarello, and D. K. Anderson. Protective effect of a 21-aminosteroid on the blood-brain barrier following subarachnoid hemorrhage in rats. *Stroke* **20,** 367–371 (1989).

CHAPTER 3

Free Radical Biology of Iron

CHRISTINE C. WINTERBOURN

ABSTRACT

This chapter describes how iron, in its various forms, is involved in redox reactions in biological systems. Some of these are normal metabolic processes, but others have the potential to initiate free radical reactions that can cause biological damage and cell death. Free radical reactions can be mediated by not only low mol wt iron complexes, but also iron-containing proteins that contain the bulk of the body's iron. Heme proteins and iron (or copper) catalyzed autoxidations are the major biological source of free radicals. The primary radicals so generated can further react with iron and hydrogen peroxide to generate hydroxyl radicals or reactive ferryl species, or to cause peroxidation of polyunsaturated lipid. Not all iron complexes are active in these processes, however. Transferrin and lactoferrin bind iron in a catalytically inactive form, and provide an important control against iron mediated radical reactions. Although iron stored intracellularly as ferritin is insulated from its surroundings by the ferritin protein shell, radical generating systems can release iron from ferritin and enable it to participate in further radical reactions.

1. INTRODUCTION

Iron is an essential constituent of a number of proteins involved in oxygen transport or metabolism. Controlled cyclic oxidation and reduction of the iron is important in these metabolic processes. However, such redox reactions can generate free radicals and other strongly oxidizing species, and if they occur in an uncontrolled manner, they can destroy the integrity of biological systems.

Free radicals can be generated from a variety of xenobiotics or metabolites. These radicals may be toxic through reacting directly with biological constituents, or they may react with O_2 to form superoxide (O_2^-) and other reactive oxygen species that may be the ultimate toxins. Reduction of oxygen to water involves the addition of four electrons. Single additions lead to the progressive formation of O_2^-, hydrogen peroxide (H_2O_2) and hydroxyl radical ($\cdot OH$). Although H_2O_2 is not a radical, it is a strong oxidant, and these intermediates of oxygen reduction are often collectively referred to as "reactive oxygen species." O_2^- itself is generally considered to have little direct toxicity, but to act as a precursor of more damaging species, e.g., H_2O_2 or $\cdot OH$ (*38,39,98*). Production of O_2^- however, is important as the initial step in the free radical chain.

By definition, free radicals contain an unpaired electron. Formation of a radical, therefore, involves a one electron oxidation or reduction of a stable species. Likewise, decay to a nonradical species involves a one electron transfer. One way of generating free radicals is through homolytic cleavage of a covalent bond. This is normally a highly energetic process requiring either elevated temperature or nuclear or UV irradiation. A more facile way of producing radicals involves electron transfer to or from a transition metal ion that can exist in two oxidation states. Iron, primarily through Fe^{2+}/Fe^{3+} transformations, is well suited to this role, and a large proportion of biological reactions that generate or remove free radicals involve iron. In some instances, this may be as a low mol wt complex, but the majority of iron-catalyzed redox reactions involve iron bound to protein. This chapter discusses how iron, in its different forms, is involved in the physiology and pathology of free radical reactions.

2. PHYSIOLOGICAL FORMS OF IRON

The majority of the iron in mammalian tissue is present as heme proteins—the oxygen transport and storage proteins hemoglobin and myoglobin, as well as various peroxidases and respiratory chain cytochromes. Nonheme iron proteins include iron–sulphur proteins e.g., xanthine oxidase and succinate dehydrogenase, as well as lipoxygenases and some oxygenases. Iron is stored intracellularly as ferritin, a protein consisting of 24 subunits that form a shell around a microcrystalline ferric oxyhydroxide core containing up to 4500 atoms of iron (*42*). Ferritin can also be deposited extracellularly, e.g., in the synovium in rheumatoid arthritis (*8,74*). Iron overload conditions lead to the accumulation of hemosiderin, a derivative of ferritin in which the protein is aggregated and partially degraded by lysosomal proteinases (*81*).

Tissue iron is highly regulated, not only because of its potential toxicity but also because of the extremely low water solubility of free Fe^{3+}. Iron is transported from the gut and in the blood bound to the plasma glycoprotein, transferrin (2). Each transferrin molecule has two high affinity binding sites for Fe^{3+}, that are normally only partially saturated. Lactoferrin is structurally similar to transferrin, but can retain its iron at a lower pH. It is present in secretions, e.g., milk and tears, and in neutrophils (2). It is released when these cells ingest microorganisms, and at sites of inflammation.

Transferrin is taken up into cells by receptor-mediated endocytosis, where it is translocated to acidic nonlysosomal vesicles (82). The low pH facilitates the removal of iron that is either incorporated into newly synthesized proteins or stored as ferritin. There is still debate about the existence of an intracellular pool of low mol wt iron in transit between these different compartments (5,8,39). Some studies suggest that there is such a pool, possibly bound to chelators, such as ADP or citrate (45,75), but definitive identification is lacking. However, this elusive pool is frequently implicated in cytotoxic free radical processes. Low mol wt iron does appear to be present extracellularly in iron overload conditions, e.g., hemochromatosis (44), but again, its identity is uncertain. Halliwell, Gutteridge, and coworkers (36,39), using a bleomycin binding assay, have measured micromolar concentrations of iron in some extracellular fluids. However, the interpretation of these results is complex (8,124), and whether the iron is low mol wt and catalytically active, is not certain.

3. BIOLOGICAL FREE RADICAL GENERATION

3.1. Enzymatic Production of Superoxide

O_2^- can be generated directly by the reaction of ferrous salts with O_2.

$$Fe^{2+} + O_2 \longrightarrow Fe^{3+} + O_2^- \tag{1}$$

Although this reaction occurs within minutes in neutral aqueous solution, it is probably of little significance biologically, because of the low concentrations of "free" iron. However, there are a number of enzymes that generate O_2^- directly. Most contain iron, which is present as heme and undergoes a reaction equivalent to (1).

The enzyme system with the greatest capacity to produce O_2^- is the NADPH oxidase of phagocytic cells. It consists of an electron transport chain that includes a heme protein, cytochrome-b_{-245}, which reacts directly with O_2 to give O_2^- (92). It is responsible for the rapid consumption of O_2 that takes place when the cells are stimulated by either particle ingestion or immunologic mediators. Intracellular NADPH is consumed, and O_2^- is released either into phagocytic vacuoles or to the outside of the cells. The O_2^- acts as a precursor of H_2O_2. In neutrophils and monocytes, H_2O_2 is converted by myeloperoxidase to hypochlorous acid. Oxidant production plays an essential role in microbial killing by the cells. However, oxidants produced in large amounts at sites of inflammation may also be damaging to host tissue (*119*).

O_2^- is produced from other heme proteins, but as a byproduct of their normal function. Hemoglobin (and myoglobin) transport O_2 as a $Fe^{2+}O_2$ complex. Although beautifully designed for this purpose, they do undergo slow autoxidation to ferrihemoglobin (or ferrimyoglobin), releasing O_2^- in the process (*68*). Approximately 3% of the total red cell hemoglobin is oxidized per day. The ferrihemoglobin is re-reduced, but the result is the continuous generation of appreciable quantities of O_2^- in the red cell (*17*). Cytochrome-P450 also produces some O_2^- as a by-product of its hydroxylation cycle (*55*).

Another important enzymatic source of O_2^- is xanthine oxidase (*62*). Xanthine oxidase is an iron-sulphur protein that also contains molybdenum and a flavin. It catalyzes the reaction of xanthine or hypoxanthine with O_2 to give uric acid and a mixture of O_2^- and H_2O_2. Recent evidence suggests that reactive oxygen species generated by xanthine oxidase contribute to the tissue injury that is associated with ischemia and reperfusion, such as occurs during a myocardial infarction (*61*). It is hypothesized that during ischemia, there is both breakdown of adenine nucleotides to hypoxanthine and conversion of xanthine dehydrogenase (the normal intracellular form of the enzyme that uses NAD rather than O_2) to the oxidase. On reoxygenation, this would allow a burst of enzymatic production of O_2^- and H_2O_2. Dramatic protective effects have been observed with superoxide dismutase and the xanthine oxidase inhibitor, allopurinol, in various reperfusion injury models.

One of the most significant metabolic sources of intracellular O_2^- is mitochondrial respiration (*30*). Although most of the O_2 consumed is converted directly to water by cytochrome oxidase, a small amount of O_2^- is produced from ubisemiquinone reacting with O_2 rather than feeding its electrons into the cytochrome chain. Iron-containing enzymes are involved in the ubiquinone reduction.

3.2. Organic Radical Production by Enzymatic Reduction

The major reductive enzymes that generate free radicals are flavoproteins, such as NADPH-cytochrome-P450 reductase, mitochondrial NADH oxidase, glutathione reductase, ferredoxin reductase, and xanthine oxidase. They reduce a broad range of substrates, many of which are toxic or are drugs with harmful side effects (Table 1). In most cases, the primary radicals reduce O_2 to O_2^-, that can then dismutate to give H_2O_2. The parent compound is regenerated, allowing continued redox cycling with concomitant production of O_2^- and H_2O_2. Much evidence has been obtained that implicates free radicals in the toxicity of these compounds (*40,60,73,86,87,109*). Iron appears to play a key role as a catalyst of secondary reactions, e.g., ·OH production or lipid peroxidation. Both reactions have been shown to accompany enzymatic reduction of a number of redox cycling compounds, provided appropriate iron salts

Table 1
Compounds that Generate Radicals on Reduction by Flavoprotein Reductases

Class	Examples	Comments
Quinones	menadione mitomycin C benzo(a)pyrine-3-6-quinone	many are cytotoxic and mutagenic
Anthracyclines	adriamycin daunomycin	anticancer agents, cardiotoxic
Quinoneimines	N-acetyl-p-benzoquinoneimine	hepatotoxic metabolite of acetaminophen
Bipyridyls	paraquat diquat	herbicides, cause lung and liver damage
Heterocyclic nitro compounds	metronidazole nitrofurantoin nifurtimox	antibacterial and antipara- sitic; mutagenic, cytotoxic and carcinogenic
Bleomycin-iron complex		active agent in bleomycin cytotoxicity

Further information on the bioreduction of these compounds can be found in refs. (50, 86).

are present (50,64,97,115,117,130). However, definitive evidence that iron is involved in their in vivo toxicity is not available. Nevertheless, recent studies showing that iron chelators can enhance the survival of mice given paraquat (54) suggest a role for chelatable iron in paraquat toxicity.

Hemoglobin is also capable of reducing many of the compounds in Table 1 to their corresponding radicals (*reviewed* in (126)). The hemoglobin is oxidized to ferrihemoglobin, and a series of radical reactions is initiated. This can lead to hemoglobin denaturation, red cell destruction, and consequent anemia.

3.3. Radical Production by Enzymatic Oxidation

Free radicals can be produced biologically by one electron oxidations catalyzed almost exclusively by heme enzymes, such as peroxidases, hemoglobin, and cytochrome P450.

3.3.1. Peroxidases

Peroxidases catalyze the generalized reaction

$$RH + H_2O_2 \longrightarrow ROH + H_2O \tag{2}$$

Table 2
Examples of Compounds that React
with Peroxidases to Generate Radicals

Class	Examples
Phenols, aminophenols arylamines	phenol
	1-naphthol
	acetaminophen
	phenetidine
	naphthylamine
	mitoxantrone
Diamino compounds	benzidine
	2-aminofluorine
Carbonyl compounds	phenylbutazone
	dihydroxyfumarate
Thiols	glutathione
	penicillamine
Polycyclic aromatics	benzo(a)pyrene

Taken from refs. (22, 37, 41, 54, 58, 79, 83, 91, 101).

This occurs, with a few exceptions, by two one-electron oxidation steps (25). Numerous compounds, including many that are cytotoxic or carcinogenic, have been shown to generate radicals by a peroxidase-catalyzed mechanism (Table 2). Most studies have been performed with horseradish peroxidase, but analagous reactions have been observed with the mammalian enzymes myeloperoxidase, eosinophil peroxidase, lactoperoxidase, thyroid peroxidase, and prostaglandin synthetase. The radicals formed may dismutate or dimerize, or react with O_2 to give either peroxy radicals or O_2^-. When O_2^- is produced, it can dismutate to give H_2O_2, allowing the reaction to self-propagate without requiring an external source of H_2O_2 (54,79).

3.3.2. Hemoglobin

Hemoglobin can oxidatively activate a wide range of drugs and xenobiotics (e.g., antimalarials, sulfonamides, acetaminophen) (126). This interaction is particularly significant in glucose-6-phosphate dehydrogenase (G6PD)-deficiency when it frequently causes red cell hemolysis (7). The drug (e.g., phenylhydrazine, $PhNHNH_2$ exemplified below) is oxidized by the heme-bound oxygen of oxyhemoglobin, to generate ferrihemoglobin, H_2O_2, and the drug radical (69,126).

$$Hb\text{-}Fe^{2+}O_2 + PhNHNH_2 + H^+ \text{ ---> } Hb^{3+} + H_2O_2 + PhN\cdot NH_2 \quad (3)$$

The drug radical reacts with O_2 to give O_2^- and a sequence of oxidative reactions with hemoglobin and other cell constituents results in denaturation and precipitation of the hemoglobin as Heinz bodies and premature removal of the cells from circulation. The susceptibility of G6PD-deficient individuals is owing to their inability to regenerate GSH required both for removing H_2O_2 (in combination with glutathione peroxidase), and for free radical scavenging.

3.3.3. Cytochrome-P450

Cytochrome-P450 is present in the endoplasmic reticulum of most mammalian cells, particularly hepatocytes. There are many isozymes that catalyze hydroxylation of different substrates by molecular oxygen (57). The mechanism involves radical intermediates, but these mostly remain associated with the enzyme. However compounds such as benzo(a)pyrene and other polycyclic hydrocarbons are oxidized by cytochrome-P450 to quinones and epoxides that undergo subsequent radical-generating reactions that contribute to their mutagenicity (58,112).

3.3.4. Lipoxygenases

Lipoxygenases are nonheme iron enzymes that insert dioxygen into polyunsaturated fatty acids to give lipid hydroperoxides (53). The products of arachidonate oxidation in mammalian systems are the inflammatory mediators, leukotrienes. Peroxidation occurs by a controlled mechanism rather than the chain reaction of lipid peroxidation described below. However, it does involve radical intermediates. These normally remain associated with the enzyme, although under certain conditions, they can be released and cause the cooxidation of antioxidants such as carotenoids (53).

3.3.5. Heme-Catalyzed Peroxidation

Peroxidases and hemoglobin can play a further role in radical-mediated biological damage, by catalyzing the reaction of H_2O_2 with essential tissue constituents. H_2O_2 is a common product of free radical processes, and an important factor in their toxicity. Fenton reactions of H_2O_2 with low mol wt iron complexes constitute one possible mechanism (*see below*), but H_2O_2 in combination with heme proteins is also strongly oxidizing. Obviously, heme proteins have essential metabolic functions, but when excessive H_2O_2 and appropriate substrates are present, peroxidases and hemoglobin can mediate peroxidation of membrane lipids (47,121) oxidative inactivation of thiol-containing enzymes or structural proteins (126), and protein crosslinking through bityrosine formation (1). It is important not to overlook these reactions in favor of Fenton reactions when considering peroxide-mediated biological damage. For example, the reaction of H_2O_2 with hemoglobin is a major contributor to drug-induced hemolysis (126).

3.4. Iron-Catalyzed Autoxidation

A number of normal metabolites, as well as xenobiotics, are oxidized by molecular oxygen. This autoxidation involves free radical intermediates and produces O_2^- and H_2O_2. Re-reduction of the oxidized compounds, e.g., by glutathione or enzymatically, sets up a redox cycle with continued production of reactive oxygen species (*49*). Compounds that readily autoxidize include thiols, hydrazines, quinols, hydroxypyrimidines, catecholamines, and compounds capable of enolization, such as ascorbic acid, δ-aminolevulinic acid, and reducing sugars (*20,27,37,59, 67,76,77,102,110,127–129*). Many of these compounds have toxic effects, to which redox cycling and radical generation almost certainly contribute. Autoxidation of glucose, for example, has been suggested to play a role in the damage to structural proteins in the lens and kidney glomerulus that occurs in uncontrolled diabetes (*132*).

Most autoxidizable compounds react only slowly with molecular oxygen, and require iron or another transition metal to catalyze the reaction. Hematin and heme proteins are good autoxidation catalysts. The efficiency of iron as a catalyst depends on its chelated form, with most chelators inhibiting oxidation. H_2O_2 is frequently involved in iron-dependent autoxidation.

Some of the more detailed autoxidation studies have been carried out on the diabetogenic agent, alloxan (and its reduced form, dialuric acid), and the fava bean constituents, divicine and isouramil, that cause red cell hemolysis in favism (*127–129*). Oxidation of these hydroxypyrimidines, once initiated, proceeds by an O_2^- dependent chain

$$RH\cdot \ + \ O_2 \ \dashrightarrow \ R \ + \ O_2^- \ + \ H^+ \qquad (4)$$

$$O_2^- \ + \ RH_2 \ + \ H^+ \ \dashrightarrow \ RH\cdot \ + \ H_2O_2 \qquad (5)$$

that proceeds rapidly and independently of a metal catalyst. Thus, even though they react slowly with O_2, their autoxidation rates are only modestly enhanced by iron. Other compounds, including hydroquinones, hydrazines, and thiols, also autoxidize mainly by an O_2^--dependent chain (*59,67,77,102*). However, in vivo where intracellular superoxide dismutase inhibits this chain, other pathways (including those involving metal ions) should become more important. Some autoxidizable compounds, e.g., sugars react poorly with O_2^-, so that most of the O_2^- formed via reaction 4 dismutates, and metal catalysis is necessary throughout the reaction.

Iron also influences autoxidations by reacting with the H_2O_2 produced, catalyzing the formation of secondary oxidants, such as ·OH, and promoting lipid peroxidation (*11*). Iron-dependent formation of ·OH (involving H_2O_2 but not O_2^-) has been observed during the autoxidation of dialuric acid (129). Autoxidizing dihydroxypyrimidines can also mediate iron-dependent lipid peroxidation (72). The mechanism does not require the O_2^- and H_2O_2 generated during autoxidation, and appears similar to that discussed below for iron and other reducing systems.

Thus, iron features in the autoxidation of biological compounds, both in initiating the process and in catalyzing subsequent potentially damaging reactions, such as ·OH production and lipid peroxidation. The bioavailability of iron catalysts of these reactions should be an important factor, therefore, in the toxicity of autoxidizable compounds. However, evidence implicating iron in the in vivo effects of compounds, such as alloxan, is still limited to protective effects of chelators being observed in some instances (19,33,43).

4. FENTON REACTIONS

The vast majority of biologically relevant free radical reactions, whether they involve autoxidizing xenobiotics or direct reduction of O_2, result in production of H_2O_2. H_2O_2 on its own is not particularly toxic, but it can react with Fe^{2+} (and other transition metal) salts to produce much more reactive species (*see reviews* (11,39,105)). These reactions are termed Fenton reactions. If Fe^{2+} can be recycled from Fe^{3+}, then the iron can act catalytically. Frequently, sufficient Fe^{2+} is present adventitiously in buffers and reagents to catalyze such reactions.

Biological damage caused by a wide variety of radical generating systems is dependent on H_2O_2 and iron, and has been attributed to Fenton reactions. A few of the many examples are: damage to DNA (14,99) and proteins, such as glutamine synthetase (56) and lens crystallins (31), enzyme inactivation (107), depolymerization of hyaluronic acid (34) and the killing of *E. coli* (90), pancreatic islets (32), and cultured hepatocytes (63,100). Asbestos is an iron-containing mineral, and Fenton chemistry may also be involved in its toxicity (120). Although readily demonstrable in experimental systems, the extent to which Fenton reactions occur physiologically is less certain, because of the difficulty of studying free radical reactions in vivo. There are now numerous reports of protection by chelators against the toxicity of free radical generating systems, suggesting that transition metal ions are involved. Some examples are given in Table 3, others are reviewed in (28,38). However, these studies must be interpreted cautiously. Usually, a source of iron was not identified, and definitive evidence that the chelators were inhibiting Fenton reactions was not obtained. Nevertheless, the accumulating circumstantial evidence, combined with our knowledge of free radical chemistry in vitro, and the recently demonstrated ability of radical generating systems to release iron from ferritin (*discussed below*), point toward a role for Fenton chemistry in free radical pathology.

Table 3
Examples of Protection by Metal Chelators Against In Vivo Toxicity

System	Chelator	Reference
Hepatotoxicity of acetaminophen	desferrioxamine	*(78)*
Paraquat toxicity	desferrioxamine & nitrilotriacetate	*(54)*
Neutrophil-mediated cytotoxicity to lungs	desferrioxamine & lactoferrin	*(111, 118)*
Joint inflammation	desferrioxamine	*(10)*
Alloxan-induced hemolysis	desferrioxamine	*(19)*

4.1. Fenton Reaction Products

The Fenton product is usually considered to be ·OH, formed according to (6).

$$(ligand)Fe^{2+} + H_2O_2 \quad ---> \quad (ligand)Fe^{3+} + \cdot OH + OH^- \quad (6)$$

Good evidence, based on product identity and quantitative scavenger studies, has been obtained for ·OH being formed both at low pH, and with Fe(EDTA) and Fe(DTPA) complexes in neutral solution (*see* discussion in $(5,11,39,105)$). The identity of the product with other iron complexes is less well established, and it appears that some can react with H_2O_2 to yield a ferryl or iron(IV) species. This can be formally represented as a ferrous-peroxo complex*, and the reaction written

$$(ligand)Fe^{2+} + H_2O_2 \quad ---> \quad (ligand)Fe^{2+}(H_2O_2) \quad (7)$$

Evidence for the formation of iron(IV) comes from kinetic studies that indicate that the Fenton product is an oxidant species with more limited reactivity than ·OH (*see* discussion in (105)). Rush and Koppenol $(94,95)$ have observed this behavior for polycarboxylate chelates that have only four or five groups that complex with the iron (compared with six for EDTA). Likewise, we have found that Fe^{2+} in neutral buffer containing no complexing agent produces an oxidant with the kinetic characteristics of iron(IV) (104). Insufficient information is available to identify the oxidant formed from Fe^{2+} chelated to physiological compounds, such as ATP, ADP or citrate, although the reactivity appears similar to that of ·OH (115). However, these compounds bind Fe^{2+} relatively weakly (compared with Fe^{3+}), and at millimolar concentrations, would still allow some Fe^{2+} to react in a nonchelated form and produce iron(IV). If iron

---

*alternative structures for an iron(IV) species include Fe^{4+}, $Fe(IV)=0$, and Fe^{2+} (H_2O_2). At present, we are not able to distinguish between these possibilities.

catalyzes Fenton reactions physiologically, therefore, the product may depend on what chelators are present, as well as their concentrations.

Hydroxyl radicals are highly reactive with almost all biological molecules. At this stage, information on the reactivity of iron(IV) complexes is limited. They should be strong oxidizing agents, although less reactive than ·OH (95,105). However, by being more selective, they could be more damaging than ·OH. It is desirable to distinguish between the two because of possible differences in reactivity with biological targets, or with scavengers and antioxidants. However, in general terms, Fenton reactions can be considered as producing a strong oxidant capable of causing biological damage regardless of its identity.

Although heme proteins react with H_2O_2 they produce little if any ·OH (88,126). In fact, the best characterized ferryl species are compound II complexes of peroxidases or the equivalent peroxide complexes of ferrous hemoglobin and myoglobin (25). These complexes are strongly oxidizing (as participants in the peroxidase cycle) but are sufficiently stable to be observed spectrally. The reaction of ferric heme with H_2O_2 could formally produce Fe(IV) and ·OH. However, the actual product is peroxidase Compound I or a ferryl hemoglobin radical (51), and little, if any, ·OH is released. Heme proteins catalyze strong oxidations, therefore, but there is little evidence that these involve ·OH.

4.2. Haber-Weiss Reaction

To achieve appreciable yields of Fenton oxidant from the low concentrations of iron that are likely to be present physiologically, the iron must be recycled. A number of radicals can do this (reaction 8), as can ascorbate (123), some thiols (93), and the cytochrome P450 system (116,122).

$$(\text{ligand})Fe^{3+} + R\cdot^- \quad ---> \quad (\text{ligand})Fe^{2+} + R \tag{8}$$

When Fe^{3+} is reduced by O_2^-, the sum of reactions 8 and 6 or 7 becomes the Haber-Weiss reaction

$$O_2^- + H_2O_2 \quad ---> \quad \cdot OH + OH^- + O_2 \tag{9}$$

This reaction does not occur in the absence of a metal catalyst, but with an appropriate catalyst, it is frequently proposed to explain the toxicity of O_2^-.

The Fe(EDTA)-catalyzed Haber Weiss reaction, which has been studied in greatest detail, proceeds efficiently with a product indistinguishable from ·OH (103,105). However, other chelates, e.g., with citrate, ATP, ADP or DTPA, are much less effective (6,103,105). Since the ferrous complexes react with H_2O_2 to give either ·OH or iron (IV) (29), it appears that the ferric complexes are poorly reduced by O_2^-. These findings imply that the Haber-Weiss reaction would not occur efficiently under most physiological conditions. Yet, there are numerous examples of biological damage caused by O_2^- and iron (*see* (*38*)), that are normally attributed to the Haber-Weiss reaction. One clue to this puzzle may be the ability of O_2^- to reduce ferritin iron (*see* below). Another relevant observation is that xanthine oxidase-derived O_2^-, possibly because it reacts in an enzyme-associated form, differs from "free" O_2^- in reacting with H_2O_2 and nonchelated iron to produce iron(IV) (*131*).

4.3. Radical-Driven Fenton Reactions

Production of ·OH or Fe(IV) accompanies the reduction of numerous redox cycling xenobiotics. The mechanism involves a radical-driven Fenton reaction. Organic radicals are generally much better than O_2^- at driving the reaction (*66,103–105,130*). For example, with paraquat, adriamycin, or anthraquinone-2-sulfonate radicals (under N_2), the more physiological chelates have similar catalytic efficiencies to Fe(EDTA) (*115*). Whether the product is ·OH or iron(IV) appears to depend on the chelator rather than the nature of the reducing radical. Radical-driven Fenton reactions, therefore, may contribute more than the Haber-Weiss reaction to biological damage by redox cycling xenobiotics. Although the reaction of the xenobiotic radicals with O_2 to give O_2^- restricts their ability to drive Fenton reactions, these reactions can still be detected aerobically. In fact, when O_2 is required as a source of H_2O_2, a peak of activity is seen at a pO_2 of 5–10 mm Hg (*130*). This implies that radical-driven Fenton reactions would be favored under physiological conditions in which the pO_2 is substantially less than that in air. These conditions prevail in most mammalian tissues.

4.4. Effects of Chelators

Chelators are frequently used to probe the involvement of iron in free radical reactions. In particular, inhibition by the relatively specific iron chelator, desferrioxamine, is often taken as proof of iron involvement (*38*). However, although iron bound to desferrioxamine is unreactive in most Fenton systems, it is not universally inert (*11*). Paraquat radicals, for example, can reduce ferrioxamine (*12*). Similarly, Fe(DTPA) catalyzes Fenton reactions driven by many organic radicals but not O_2^- (*6,103,130*). Also of note is that chelators, like other organic molecules, are good ·OH scavengers. Inhibition of a reaction by high concentrations of chelator could reflect this mechanism. Desferrioxamine, for example, can react with various radicals, including O_2^- (*23*), myeloperoxidase-derived oxidants (*52*), and intermediates of hemoglobin-catalyzed lipid peroxidation (*48*).

4.5. Site Specific Reactions

An important consideration for biological free radical reactions is compartmentalization. Being highly reactive, free radicals tend to react close to the site of generation. Consequently, constituents of one compartment may not influence reactions of radicals generated in another (*11,21*). Likewise, the breakdown of compartmentalization may allow the mixing of separated reactants so that radical reactions can proceed. Free radical production, therefore, could be a consequence of injury by another mechanism.

Permeability and solubility considerations lead to distinctions, not only between cellular and extracellular compartments, but also, within organelles and between lipid membranes and the aqueous phase. Thus, extracellular superoxide dismutase scavenges very little O_2^- generated intracellularly, and the protection against H_2O_2 cytotoxicity by only those chelators that penetrate the cell membrane (*99*) suggests involvement of intracellularly located metal ions.

The term "site specific" is also used to describe a reaction occurring at a particular site on a macromolecule or biological structure (*11,107*), such as when iron or copper binds to a protein and initiates a reaction at the binding site (*11,18,21,56*). A general feature is that H_2O_2 and metal ions are involved, but the reaction is not inhibited by ·OH scavengers. This is explained on the basis of limited access to the reaction site. O_2^- or other radicals may be involved in reducing the iron. These reactions are frequently referred to as site specific ·OH generation. However, by definition, scavengers do not give the expected inhibition for a reaction of ·OH, and whether the product is ·OH or iron (IV), or involves a concerted reaction with the target, is almost impossible to establish (*105*). A preferable term, therefore, is site specific Fenton reaction.

Although lack of inhibition by ·OH scavengers may be characteristic of site specific Fenton reactions, anomalous scavenger behavior could also indicate a homogenous reaction involving iron(IV) rather than ·OH (*105*). Iron binding to the target should be demonstrated before a site specific mechanism is proposed. Note, however, that inhibition of the detection reaction by a chelator may not necessarily show this, since chelators can also change the Fenton product, as described above.

5. LIPID PEROXIDATION

Polyunsaturated fatty acids are essential constituents of biological membranes. They are prone to oxidation by molecular oxygen by a radical chain process called lipid peroxidation. Peroxidation can be initiated by ionizing or UV irradiation (e.g., butter going rancid in sunlight) or by reactions mediated by iron or other transition metals. Iron can also

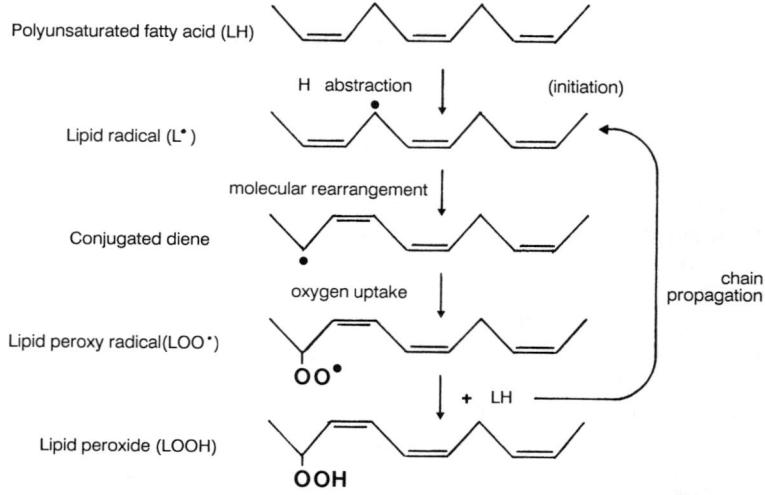

Fig. 1.   Initiation and propagation of lipid peroxidation.

enhance peroxidation by catalyzing the breakdown of lipid peroxides, leading to secondary chain reactions (for more detailed reviews, *see* (5,38,46)). Low mol wt iron complexes and heme derivatives catalyze these reactions. The reactions involved in initiation and propagation of the chain are shown in Fig. 1. Lipid peroxides are not very stable, and break down to products that include malondialdehyde, other aldehydes, and hydrocarbon gases (26). True initiation involves hydrogen abstraction. ·OH radicals are capable of this, and do so in radiation-induced peroxidation. However, ·OH appears to play a minor role in lipid peroxidation catalyzed by iron salts. Although the mechanism is not yet fully understood, it seems that true initiation contributes little to iron-catalyzed peroxidation, which occurs mostly as a result of breakdown of preexisting hydroperoxides (5,65). These are almost inevitably present in isolated lipid preparations. Only minute concentrations, which may not even be measurable, are needed for this mechanism to predominate. Even if there is some true initiation, it is probably soon overwhelmed by secondary reactions.

Fe^{2+} can decompose lipid peroxides (24):

$$LOOH + Fe^{2+} \longrightarrow LO\cdot + Fe^{3+} + OH^- \qquad (10)$$

The equivalent reaction of Fe^{3+}

$$LOOH + Fe^{3+} \longrightarrow LOO\cdot + Fe^{2+} + H^+ \qquad (11)$$

appears to be less facile, although Fe^{3+} is active in some peroxidation systems (5,38). The alkoxy and peroxy radicals formed can enter the propagation step of the chain by abstracting a hydrogen atom as in Fig. 1. However, iron-dependent lipid peroxidation appears to be more complex than this. It requires a reductant for maximum effect. O_2^- and other reducing radicals, as well as the microsomal cytochrome P450 system and ascorbate, are effective (5,65,112,117). This may indicate involvement of a perferryl species ($Fe^{2+}O_2$) (113). Catalase is generally inhibitory with heme compounds but not with low mol wt iron complexes (5,47,121). Many chelators, including EDTA, inhibit peroxidation, but Fe(ADP) and Fe(ATP) are active catalysts. Thus, Fenton production of ·OH (or iron (IV)) does not appear to be an important step in the mechanism. Work from the laboratories of Aust (5,65) and Braughler (13) has shown that in various systems, lipid peroxidation is most effective with equimolar concentrations of Fe^{2+} and Fe^{3+}. They propose that a complex of Fe^{2+}, Fe^{3+}, and O_2 is involved in the process. Another possibility is that, because lipid peroxidation is a complex multistep reaction, Fe^{2+} and Fe^{3+} might each influence a different step in the chain. It should also be remembered that most detection methods, e.g., the commonly used thiobarbituric acid assay, measure only a minor product of lipid peroxidation, changes to which might not always reflect changes to the main peroxidation pathway.

Even though aspects of the mechanism of lipid peroxidation are not yet understood, it is clear that iron plays a major role in the process. It is also apparent that the requirements for iron-dependent lipid peroxidation are different from those for Fenton chemistry. Iron complexes show different reactivities in the two systems, and only in the latter is H_2O_2 an absolute requirement. It seems likely, therefore, that physiological situations will occur where one but not the other reaction is favored.

6. TRANSFERRIN AND LACTOFERRIN

The iron binding proteins, transferrin and lactoferrin, are important, not only for transporting iron in a soluble form, but also for protecting against iron-mediated oxidative reactions. Although an early report (3) proposed that lactoferrin could catalyze the Haber-Weiss reaction and enable neutrophils to produce ·OH, this has subsequently been discounted (4,125). It is now generally accepted that iron bound to transferrin and lactoferrin is ineffective at catalyzing either Fenton oxidations or lipid peroxidation (15). Thus, adding the apoproteins to systems containing adventitious or added iron salts inhibits both processes (4,35).

Physiologically, iron released into plasma or extracellular fluids should be taken up by transferrin, and not pose an oxidative threat until the binding capacity is exceeded. Lactoferrin and transferrin do not take up heme iron or inhibit heme-dependent reactions. Release of hemo-globin into plasma, therefore, could cause problems. However, hapto-globin can restrict the oxidative reactions of hemoglobin, and may be protective (85). High concentrations of H_2O_2 can release some iron from hemoglobin by destroying the heme group (88). Transferrin or desfer-rioxamine can bind this iron and partially inhibit accompanying oxida-tions.

Lactoferrin in milk and neutrophils has a low iron loading (114). By taking up iron, it can inhibit bacterial growth, and may also protect against oxidative reactions of neutrophils. It can take up iron released from ferritin by O_2^- (71), and may take up iron released from degraded tissue or bacteria.

Although transferrin acts as an antioxidant at neutral pH, it starts to lose its iron at about pH 5, and becomes a Fenton or lipid peroxidation catalyst (6,96). Iron release is a prerequisite for this activity. Thus, trans-ferrin iron has the potential to act as a prooxidant in low pH environ-ments, such as in lysosomes. In fact, lysosomes might be one of the more critical sites of lipid peroxidation within the cell, since a resultant increase in membrane permeability would allow secondary damage by released lysosomal hydrolases. Lactoferrin iron, on the other hand, is retained to a lower pH, and should not be released when exposed to acid conditions in the neutrophil phagosome or at sites of inflammation.

7. FERRITIN

Although physiological pools of low mol wt iron remain elusive, impetus to the argument that iron-mediated oxidative reactions are of physiological importance comes from the recent demonstration that a variety of radical-generating systems can release iron from ferritin (8). The iron in the ferritin core is insulated from its surroundings, and requires reduction and/or chelation for its release. This occurs only slow-ly with physiological reductants, e.g., ascorbate, but a number of free radical generating systems are much more effective. O_2^- can mediate the release (8,9), and paraquat, adriamycin, and some semiquinone and nitroaromatic radicals, provided their redox potentials are below about -0.24 V, are even more efficient (70,106,107). Diquat administered to mice has recently been shown to release iron from liver ferritin (89). The cytotoxic xenobiotics 6-hydroxydopamine, dialuric acid, divicine, and isouramil give very rapid release of iron from ferritin (72). Ferritin, therefore, must be considered as a potential iron source for catalyzing a wide variety of free radical reactions.

Iron released from ferritin can catalyze further autoxidation or partic-ipate in subsequent ·OH production or lipid peroxidation (4,7,16,72,81,108,117). It is, however, subject to chelation, so that desfer-rioxamine and EDTA inhibit ferritin-dependent lipid peroxidation

(*4,117*). The physiological reactivity of released ferritin iron will also depend on what it complexes with, and as described above, this will determine whether it can produce ·OH or mediate lipid peroxidation. The demonstration that iron released from ferritin by O_2^- is effectively transferred to transferrin or lactoferrin (*71*) infers that such iron will not necessarily be available to participate in further radical reactions. Thus, transferrin should protect against the effects of ferritin iron released at extracellular sites. Likewise, although iron can be released from ferritin by stimulated neutrophils (*9*), its prooxidant activity could be restricted by coreleased lactoferrin.

Hemosiderin iron has some prooxidant activity in mediating lipid peroxidation and ·OH production, but at neutral pH, it is considerably less effective than ferritin (*81*). However, at lower pH, hemosiderin releases some iron and becomes a more effective prooxidant (*84*). Therefore, hemosiderin-dependent oxidative damage may be important in lysosomes, where the majority of this protein is located. Increased lysosomal fragility is associated with increased hemosiderin levels (*81*).

8. NUTRITIONAL CONSIDERATIONS

Although iron is an essential element for a variety of metabolic functions, its ability to participate in free radical reactions implies that dietary sufficiency, and not excess, is desirable. However, since essential iron, such as in heme proteins, can generate radicals, iron-dependent radical generation cannot be prevented by dietary regulation. Control depends on the body's normal antioxidant defense. There is sound rationale for balancing high iron intake with increased levels of antioxidants, e.g., vitamins C, E, and A and, possibly, selenium. However, some caution is needed, particularly with vitamin C. Iron/ascorbate can act as a prooxidant, and high doses of ascorbate in association with iron overload have had undesirable effects (*80*).

REFERENCES

1. R. Aeschbach, R. Amado, and H. Neukom. Formation of dityrosine cross-links in proteins by oxidation of tyrosine residues. *Biochim. Biophys. Acta* **439**, 292–301 (1976).
2. P. Aisen, and I. Listowski. Iron transport and storage proteins. *Ann. Rev. Biochem.* **49**, 357–393 (1980).
3. D. R. Ambruso, and R. B. Johnston. Lactoferrin enhances hydroxyl radical production by human neutrophils, neutrophil particulate fractions, and an enzymatic generating system. *J. Clin. Invest.* **67**, 352–360 (1981).
4. O. I. Aruoma, and B. Halliwell. Superoxide-dependent and ascorbate-dependent formation of hydroxyl radicals from hydrogen peroxide in the presence of iron. Are lactoferrin and transferrin promoters of hydroxyl-radical generation? *Biochem. J.* **241**, 273–278 (1987).

5. S. D. Aust, L. A. Morehouse, and C. E. Thomas. Role of metals in oxygen radical reactions. *J. Free Rad. Biol. Med.* **1**, 3–25 (1985).
6. M. S. Baker, and J. M. Gebicki. The effect of pH on yields of hydroxyl radicals produced from superoxide by potential biological iron chelators. *Arch. Biochem. Biophys.* **246**, 581–588 (1986).
7. E. Beutler. Glucose-6-phosphate dehydrogenase deficiency. In *The Metabolic Basis of Inherited Disease*, J. B. Stanbury, J. B. Wyngaarden, D. S. Fredrickson, J. L. Goldstein, and M. S. Brown, eds., McGraw Hill, New York, 1983, pp. 1629–1653.
8. P. Biemond, A. J. G. Swaak, H. G. Van Eijk, and J. F. Koster. Superoxide dependent iron release from ferritin in inflammatory diseases. *Free Rad. Biol. Med.* **4**, 185–198 (1988).
9. P. Biemond, H. G. Van Eijk, A. J. G. Swaak, and J. F. Koster. Iron mobilization from ferritin by superoxide derived from stimulated polymorphonuclear leucocytes. Possible mechanism in inflammation diseases. *J. Clin. Invest.* **73**, 1576–1579 (1984).
10. D. R. Blake, N. D. Hall, P. A. Bacon, P. A. Dieppe, B. Halliwell, and J. M. C. Gutteridge. Effect of a specific iron-chelating agent on animal models of inflammation. *Ann. Rheum. Dis.* **42**, 89–93 (1983).
11. D. C. Borg, and K. M. Schaich. Cytotoxicity from coupled redox cycling of autoxidizing xenobiotics and metals. *Israel J. Chem.* **24**, 38-53 (1984).
12. D. C. Borg, and K. M. Schaich. Prooxidant action of desferrioxamine: Fenton-like production of hydroxyl radicals by reduced ferrioxamine. *J. Free Rad. Biol. Med.* **2**, 237–243 (1986).
13. J. M. Braughler, R. L. Chase, and J. F. Pregenzer. Oxidation of ferrous iron during peroxidation of lipid substrates. *Biochim. Biophys. Acta* **921**, 457–464 (1987).
14. K. Brawn, and I. Fridovich. DNA strand scission by enzymically generated oxygen radicals. *Arch. Biochem. Biophys.* **206**, 414–419 (1981).
15. B. E. Britigan, M. S. Cohen, and G. M. Rosen. Detection of the production of oxygen-centered free radicals by human neutrophils using spin trapping techniques: a critical perspective. *J. Leukocyte Biol.* **41**, 349–362 (1987).
16. G. Carlin, and K. E. Arfors. Peroxidation of liposomes promoted by human polymorphonuclear leucocytes. *J. Free Rad. Biol. Med.* **1**, 437–442 (1985).
17. R. W. Carrell, C. C. Winterbourn, and E. A. Rachmilewitz. Activated oxygen and haemolysis. *Br. J. Haematol.* **30**, 259–264 (1975).
18. M. Chevion. A site-specific mechanism for free radical induced biological damage: the essential role of redox-active transition metals. *Free Rad. Biol. Med.* **5**, 27–37 (1988).
19. I. A. Clark, and N. H. Hunt. Evidence for reactive oxygen intermediates causing hemolysis and parasite death in malaria. *Infect. Immun.* **39**, 1–6 (1983).
20. G. Cohen, and R. E. Heikkila. The generation of hydrogen peroxide, superoxide radical, and hydroxyl radical by 6-hydroxydopamine, dialuric acid, and related cytotoxic agents. *J. Biol. Chem.* **249**, 2447–2452 (1974).
21. G. Czapski. On the use of OH· scavengers in biological systems. *Israel J. Chem.* **24**, 29–32 (1984).
22. M. D'Arcy Doherty, I. Wilson, P. Wardman, J. Basra, L. H. Patterson and G. M. Cohen. Peroxidase activation of 1-naphthol to naphthoxy or naphthoxy-derived radicals and their reaction with glutathione. *Chem. Biol. Interact.* **58**, 199–215 (1986).

23. M. J. Davies, R. Donkor, C. A. Dunster, C. A. Gee, S. Jonas, and R. L. Willson. Desferrioxamine (desferal) and superoxide free radicals. Formation of an enzyme-damaging nitroxide *Biochem. J.* **246,** 725–729 (1987).
24. M. J. Davies, and T. F. Slater. Studies on the metal-ion and lipoxygenase-catalyzed breakdown of hydroperoxides using electron-spin-resonance spectroscopy. *Biochem. J.* **245,** 167–173 (1987).
25. H. B. Dunford. Free Radicals in iron-containing systems. *Free Rad. Biol. Med.* **3,** 405–421 (1987).
26. H. Esterbauer. Lipid peroxidation products: formation, chemical properties, and biological activities. In *Free Radicals in Liver Injury,* G. Poli, K. H. Cheeseman, M. U. Dianzani, and T. F. Slater, eds., IRL Press, Oxford, 1985, pp. 29–47.
27. P. Eyer, and E. Longfelder. Radical formation during autoxidation of 4-dimethylaminophenol and some properties of the reaction products. *Biochem. Pharmacol.* **33,** 1005–1013 (1984).
28. J. C. Fantone, and P. A. Ward. Role of oxygen derived free radicals and metabolites in leucocyte-dependent inflammatory reactions. *Am. J. Pathol.* **107,** 397–418 (1982).
29. R. A. Floyd and C. A. Lewis. Hydroxyl free radical formation from hydrogen peroxide by ferrous iron-nucleotide complexes. *Biochemistry* **22,** 2645–2649 (1983).
30. H. J. Forman, and A. Boveris. Superoxide radical and hydrogen peroxide in mitochondria. In *Free Radicals in Biology,* W. A. Pryor, ed., Academic, New York, 1982, pp. 65–90.
31. D. Garland, J. S. Zigler, and J. Kinoshita. Structural changes in bovine lens crystallins induced by ascorbate, metal, and oxygen. *Arch. Biochem. Biophys.* **251,** 771–776 (1986).
32. K. Grankvist, S. Marklund, J. Sehlin, and I.-.B. Taljedal. Superoxide dismutase, catalase, and scavengers of hydroxyl radical protect against the toxic action of alloxan on pancreatic islet cells. *Biochem. J.* **182,** 17–25 (1979).
33. K. Grankvist, S. Marklund, and I. B. Taljedal. Influence of trace metals on alloxan cytotoxicity in pancreatic islets. *FEBS Lett.* **105,** 15–18 (1979).
34. R. A. Greenwald, and W. W. Moy. Effect of oxygen-derived free radicals on hyaluronic acid. *Arthritis Rheum.* **23,** 455–463 (1980).
35. J. M. C. Gutteridge, S. K. Paterson, A. W. Segal, and B. Halliwell. Inhibition of lipid peroxidation by the iron-binding protein lactoferrin. *Biochem. J.* **199,** 259–261 (1981).
36. J. M. C. Gutteridge, D. A. Rowley, and B. Halliwell. Superoxide-dependent formation of hydroxyl radicals in the presence of iron salts. Detection of "free" iron in biological systems by using bleomycin-dependent degradation of DNA. *Biochem. J.* **199,** 263–265 (1981).
37. B. Halliwell. Generation of hydrogen peroxide, superoxide, and hydroxyl radicals during the oxidation of dihydroxyfumaric acid by peroxidase. *Biochem. J.* **163,** 441–448 (1977).
38. B. Halliwell, and J. M. C. Gutteridge. The importance of free radicals and catalytic metal ions in human diseases. *Mol. Aspects Med.* **8,** 89–193 (1985).
39. B. Halliwell, and J. M. C. Gutteridge. Oxygen toxicity, oxygen radicals, transition metals, and disease. *Biochem. J.* **219,** 1–14 (1984).
40. K. Handa, and S. Sato. Generation of free radicals of quinone group-containing anticancer chemicals in NADPH-microsome system as evidenced by initiation of sulfite oxidation. *Gann* **66,** 43–47 (1975).

41. L. S. Harman, D. K. Carver, J. Schreiber, and R. P. Mason. One- and two-electron oxidation of reduced glutathione by peroxidases. *J. Biol. Chem.* **261,** 1642–1648 (1986).
42. P. M. Harrison. Ferritin: An iron-storage molecule. *Sem. Hematol.* **14,** 55–70 (1977).
43. R. E. Heikkila, and F. S. Cabbat. The prevention of alloxan-induced diabetes in mice by the iron-chelator detapac: suggestion of a role for iron in the cytotoxic process. *Experientia* **38,** 378–379 (1982).
44. C. Hershko, and T. E. A. Peto. Non-transferrin plasma iron. *Br. J. Hematol.* **66,** 149–151 (1987).
45. A. Jacobs. Low molecular weight intracellular iron transport compounds. *Blood* **50,** 433–439 (1977).
46. J. Kanner, J. B. German, and J. E. Kinsella. Initiation of lipid peroxidation in biological systems. *CRC Crit. Rev. Food Sci. Nutr.* **25,** 317–364 (1987).
47. J. Kanner, and S. Harel. Initiation of membranal lipid peroxidation by activated metmyoglobin and methemoglobin. *Arch. Biochem. Biophys.* **237,** 314–321 (1985).
48. J. Kanner, and S. Harel. Desferrioxamine as an electron donor. Inhibition of membranal lipid peroxidation initiated by H_2O_2-activated metmyoglobin and other peroxidizing systems. *Free Rad. Res. Commun.* **3,** 309–317 (1987).
49. H. Kappus and H. Sies. Toxic drug effects associated with oxygen metabolism: redox cycling and lipid peroxidation. *Experientia* **37,** 1233–1241 (1981).
50. H. Kappus. Overview of enzyme systems involved in bioreduction of drugs and in redox cycling. *Biochem. Pharmacol.* **35,** 1–6 (1986).
51. N. K. King, F. D. Looney, and M. E. Winfield. Amino acid free radicals in oxidized metmyoglobin. *Biochim. Biophys. Acta* **113,** 65–82 (1976).
52. S. J. Klebanoff, and A. M. Watersdorf. Inhibition of peroxidase-catalyzed reactions by deferoxamine. *Arch. Biochem. Biophys.* **264,** 600–606 (1988).
53. B. P. Klein, D. King, and S. Grossman. Cooxidation reactions of lipoxygenase in plant systems. *Adv. Free Rad. Biol. Med.* **1,** 309–343 (1985).
54. R. Kohen, and M. Chevion. Paraquat toxicity is enhanced by iron and inhibited by desferrioxamine in laboratory mice. *Biochem. Pharmacol.* **34,** 1841–1843 (1985).
55. H. Kuthan, H. Tsuji, H. Graf, V. Ullrich, J. Werringloer, and R. W. Estabrook. Generation of superoxide as a source of hydrogen peroxide in a reconstituted monooxygenase system. *FEBS Lett.* **91,** 343–345 (1978).
56. R. L. Levine. Oxidative Modification of Glutamine Synthetase. II. Characterization of the Ascorbate Model System. *J. Biol. Chem.* **258,** 11828–11833 (1983).
57. B. G. Malstrom. Enzymology of oxygen. *Annu. Rev. Biochem.* **51,** 21–59 (1982).
58. L. J. Marnett, and T. E. Eling. Cooxidation during prostaglandin synthesis: a pathway for the metabolic activation of xenobiotics. In *Reviews of Biochemical Toxicology*, E. Hodgson, J. R. Bend, and R. M. Philpot, eds., Elsevier, New York, 1983, pp. 135–172.
59. T. Mashino, and I. Fridovich. Superoxide radical initiates the autoxidation of dihydroxyacetone. *Arch. Biochem. Biophys.* **254,** 547–551 (1987).
60. R. P. Mason (ed.). Free Radical Metabolites of Toxic Chemicals. *Environ. Hlth. Perspect.* **64** (1985).
61. J. M. McCord. Oxygen-derived free radicals in postischemic tissue injury. *N. Engl. J. Med.* **312,** 159–163 (1985).
62. J. M. McCord, and I. Fridovich. The reduction of cytochrome-*c* by milk xanthine oxidase. *J. Biol. Chem.* **243,** 5753–5760 (1968).

63. A. C. Mello Filho, M. E. Hoffmann, and R. Meneghini. Cell killing and DNA damage by hydrogen peroxide are mediated by intracellular iron. *Biochem. J.* **218,** 273–275 (1984).

64. E. G. Mimnaugh, M. A. Trush, and T. E. Gram. Stimulation by adriamycin of rat heart and liver microsomal NADPH-dependent lipid peroxidation. *Biochem. Pharmacol.* **30,** 2797–2804 (1981).

65. G. Minotti, and S. D. Aust. The requirement for iron(III) in the initiation of lipid peroxidation by iron(II) and hydrogen peroxide. *J. Biol. Chem.* **262,** 1098–1104 (1987).

66. D. Mira, U. Brunk, A. Boveris, and E. Cadenas. One-electron transfer reactions of diquat radical to different reduction intermediates of oxygen. *Free Rad. Biol. Med.* **5,** 155–163 (1988).

67. H. P. Misra, and I. Fridovich. The role of superoxide anion in the autoxidation of epinephrine and a simple assay for superoxide dismutase. *J. Biol. Chem.* **247,** 3170–3180 (1972).

68. H. P. Misra, and I. Fridovich. The generation of superoxide radical during the autoxidation of hemoglobin. *J. Biol. Chem.* **247,** 6960–6962 (1972).

69. H. P. Misra, and I. Fridovich. The oxidation of phenylhydrazine: Superoxide and mechanism. *Biochemistry* **15,** 681–687 (1976).

70. H. P. Monteiro, G. F. Vile, and C. C. Winterbourn. Release of iron from ferritin by semiquinone, anthracycline, bipyridyl, and nitroaromatic radicals. *Free Rad. Biol. Med.* **6,** 587–591, (1989).

71. H. P. Monteiro, and C. C. Winterbourn. The superoxide-dependent transfer of iron from ferritin to transferrin and lactoferrin. *Biochem. J.* **256,** 923–928 (1989).

72. H. P. Monteiro, and C. C. Winterbourn. Release of iron from ferritin by divicine, isouramil, acid-hydrolyzed vicine and dialuric acid, and initiation of lipid peroxidation. *Arch. Biochem. Biophys.* **271,** 536–545 (1989).

73. S. N. J. Moreno, and R. Docampo. Mechanism of toxicity of nitro compounds used in the chemotherapy of trichomoniasis. *Environ. Hlth. Perspec.* **64,** 199–208 (1985).

74. K. D. Muirden. The anemia of rheumatoid arthritis. The significance of iron deposits in the synovial membrane. *Aust. Ann. Med.* **2,** 97–104 (1970).

75. M. Mulligan, B. Althaus, and M. C. Linder. Nonferritin, nonheme iron pools in rat tissues. *Int. J. Biochem.* **18,** 791–798 (1986).

76. R. Munday. Generation of superoxide radical and hydrogen peroxide by 1,2,4-triaminobenzene, a mutagenic and myotoxic aromatic amine. *Chem. Biol. Interact.* **60,** 171–181 (1986).

77. R. Munday. Generation of superoxide radical, hydrogen peroxide, and hydroxyl radical during the autoxidation of N,N,N', N'-tetramethyl-p-phenylenediamine. *Chem. Biol. Interact.* **65,** 133–143 (1988).

78. D. Nakai, J. W. Oakes, and J. L. Farber. Potentiation in the intact rat of the hepatotoxicity of acetaminophen by 1,3-bis(2-chloroethyl)-1-nitrosourea. *Arch. Biochem. Biophys.* **267,** 651–659 (1988).

79. M. Nakamura, I. Yamazaki, S. Ohtaki, and S. Nakamura. Characterization of one- and two-electron oxidations of glutathione, coupled with lactoperoxidase and thyroid peroxidase reactions. *J. Biol. Chem.* **261,** 13923–13927 (1986).

80. A. W. Nienhuis. Vitamin C and iron. *N. Engl. J. Med.* **304,** 170–171 (1981).

81. M. O'Connell, B. Halliwell, C. P. Moorhouse, O. I. Aruoma, H. Baum, and T. J. Peters. Formation of hydroxyl radicals in the presence of ferritin and hemosiderin. Is hemosiderin formation a biological protective mechanism? *Biochem. J.* **234,** 727–731 (1986).

82. J. N. Octave, Y. J. Schneider, A. Trouet, and R. R. Crichton. Iron uptake and utilization by mammalian cells. I: Cellular uptake of transferrin and iron. *Trends Biochem. Sci.* **8**, 217–220 (1983).

83. P. Ortiz de Montellano and L. A. Grab. Cooxidation of styrene by horseradish peroxidase and glutathione. *Mol. Pharmacol.* **30**, 666–669 (1986).

84. M. Ozaki, T. Kawabata, and M. Awai. Iron release from hemosiderin and production of iron-catalyzed hydroxyl radicals in vitro. *Biochem. J.* **250**, 589–595 (1988).

85. S. S. Panter, S. M. Sadrzadeh, P. E. Hallaway, J. Haines, V. E. Anderson, and J. W. Eaton. Hypohaptoglobinemia associated with familial epilepsy. *J. Exp. Med.* **161**, 748–754 (1985).

86. G. Powis. Free radical formation by antitumor quinones. *Free Rad. Biol. Med.* **6**, 63–101 (1989).

87. G. Powis, B. A. Svingen, and P. Appel. Quinone-stimulated superoxide formation by subcellular fractions, isolated hepatocytes, and other cells. *Mol. Pharmacol.* **20**, 387–394 (1981).

88. A. Puppo and B. Halliwell. Formation of hydroxyl radicals from hydrogen peroxide in the presence of iron. Is hemoglobin a biological Fenton catalyst? *Biochem. J.* **249**, 185–190.

89. D. W. Reif, I. L. P. Beales, C. E. Thomas, and S. D. Aust. Effect of diquat on the distribution of iron in rat liver. *Toxicol. Appl. Pharmacol.* **93**, 506–510 (1988).

90. J. E. Repine, R. B. Fox, and E. M. Berger. Hydrogen peroxide kills *Staphylococcus aureus* by reacting with staphylococcal iron to form hydroxyl radical. *J. Biol. Chem.* **256**, 7094–7096 (1981).

91. D. Ross, and P. Moldeus. Generation of reactive species and fate of thiols during peroxidase-catalyzed metabolic activation of aromatic amines and phenols. *Environ. Hlth. Perspect.* **64**, 253–257 (1985).

92. F. Rossi. The O_2^--forming NADPH oxidase of the phagocytes: nature, mechanisms of activation and function. *Biochim. Biophys. Acta* **853**, 65–89 (1986).

93. D. A. Rowley, and B. Halliwell. Superoxide-dependent formation of hydroxyl radicals in the presence of thiol compounds. *FEBS Lett.* **138**, 33–36 (1982).

94. J. D. Rush and W. H. Koppenol. The reaction between ferrous polyaminocarboxylate complexes and hydrogen peroxide: An investigation of the reaction intermediates by stopped flow spectrophotometry. *J. Inorg. Biochem.* **29**, 199–215 (1987).

95. J. D. Rush and W. H. Koppenol. Reactions of FeIInta and FeIIedda with hydrogen peroxide. *J. Amer. Chem. Soc.* **110**, 4957–4963 (1988).

96. M. Saito, L. A. Morehouse, and S. D. Aust. Transferrin-dependent lipid peroxidation. *J. Free Rad. Biol. Med.* **2**, 99–105.

97. M. S. Sandy, P. Moldeus, D. Ross, and M. T. Smith. Role of redox cycling and lipid peroxidation in bipyridyl herbicide cytotoxicity. *Biochem. Pharmacol.* **35**, 3095–3101 (1986).

98. D. T. Sawyer, and J. S. Valentine. How super is superoxide? *Acc. Chem. Res.* **14**, 393–400 (1981).

99. I. Schraufstatter, P. A. Hyslop, J. H. Jackson, and C. G. Cochrane. Oxidant-induced DNA Damage of Target Cells. *J. Clin. Invest.* **82**, 1040–1050 (1988).

100. P. E. Starke and J. L. Farber. Ferric iron and superoxide are required for the killing of cultured hepatocytes by hydrogen peroxide. *J. Biol. Chem.* **260**, 10099–10104 (1985).

101. V. V. Subrahamanyam and P. J. O'Brien. Peroxidase-catalyzed oxygen activation by arylamine carcinogens and phenol. *Chem. Biol. Interact.* **56,** 185–199 (1985).
102. S. G. Sullivan and A. Stern. Effects of superoxide dismutase and catalase on catalysis of 6-hydroxydopamine and 6-aminodopamine autoxidation by iron and ascorbate. *Biochem. Pharmacol.* **30,** 2279–2285 (1981).
103. H. C. Sutton. Efficiency of chelated iron compounds as catalysts for the Haber-Weiss reaction. *J. Free Rad. Biol. Med.* **1,** 195–202 (1985).
104. H. C. Sutton, G. F. Vile, and C. C. Winterbourn. Radical driven Fenton reactions—Evidence from paraquat radical studies for production of tetravalent iron in the presence and absence of EDTA. *Arch. Biochem. Biophys.* **256,** 462–471 (1987).
105. H. C. Sutton and C. C. Winterbourn. On the participation of higher oxidation states of iron and copper in Fenton-type reactions. *Free Radical Biol. Med.* **6,** 53–60 (1989).
106. C. E. Thomas and S. D. Aust. Reductive release of iron from ferritin by cation free radicals of paraquat and other bipyridyls. *J. Biol. Chem.* **261,** 13064–13070 (1986).
107. C. E. Thomas and S. D. Aust. Release of iron from ferritin by cardiotoxic anthracycline antibiotics. *Arch. Biochem. Biophys.* **248,** 684–689 (1986).
108. C. E. Thomas, L. A. Morehouse, and S. D. Aust. Ferritin and superoxide-dependent lipid peroxidation. *J. Biol. Chem.* **260,** 3275–3280 (1985).
109. H. Thor, M. T. Smith, P. Hartzell, G. Bellomo, S. Jewell, and S. Orrenius. The metabolism of menadione (2-methyl-1,4-naphthoquinone) by isolated hepatocytes. *J. Biol. Chem.* **257,** 12419–12425 (1982).
110. P. Thornalley, S. Wolff, J. Crabbe, and A. Stern. The autoxidation of glyceraldehyde and other simple monosaccharides under physiological conditions catalyzed by buffer ions. *Biochim. Biophys. Acta* **797,** 276–287 (1984).
111. G. O. Till, J. R. Hatherill, W. W. Tourtellotte, M. J. Lutz, and P. A. Ward. Lipid peroxidation and acute lung injury after thermal trauma to skin. Evidence of a role for hydroxyl radical. *Am. J. Pathol.* **119,** 376–384 (1985).
112. P. O. P. T'so, W. J. Caspary, and R. J. Lorentzen. The involvement of free radicals in chemical carcinogenesis. In *Free Rad. Biol.,* W. A. Pryor, ed., vol. 3, Academic, New York, 1977, pp. 251–303.
113. F. Ursini, M. Maiorino, P. Hochstein, and L. Ernster. Microsomal lipid peroxidation: mechanisms of initiation. The role of iron and iron chelates. *Free Rad. Biol. Med.* **6,** 31–36 (1989).
114. J. J. Van Snick, P. L. Masson, J. F. Heremans. The involvement of lactoferrin in the hyposideremia of acute inflammation. *J. Exp. Med.* **140,** 1068–1084 (1974).
115. G. Vile, C. C. Winterbourn, and H. C. Sutton. Radical-driven Fenton reactions: Studies with paraquat, adriamycin, and anthraquinone-2-sulphonate, and with citrate, ATP, ADP, and pyrophosphate iron chelates. *Arch. Biochem. Biophys.* **259,** 616–626 (1987).
116. G. F. Vile and C. C. Winterbourn. Microsomal reduction of low molecular weight Fe^{3+}-chelates and ferritin: enhancement by adriamycin, paraquat, menadione and anthraquinone-2-sulphonate and inhibition by oxygen. *Arch. Biochem. Biophys.* **267,** 606–613 (1988).
117. G. F. Vile and C. C. Winterbourn. Adriamycin-dependent peroxidation of rat liver and heart microsomes catalyzed by iron chelates and ferritin: maximum peroxidation at low oxygen partial pressures. *Biochem. Pharmacol.* **37,** 2893–2897 (1988).

118. P. A. Ward, G. O. Till, R. Kunkel, and C. Beauchamp. Evidence for role of hydroxyl radical in complement and neutrophil-dependent tissue injury. *J. Clin. Invest.* **72,** 789–801 (1983).

119. S. J. Weiss and A. F. LoBuglio. Phagocyte-generated oxygen metabolites and cellular injury. *Lab. Invest.* **47,** 5–18 (1982).

120. S. A. Weitzman, and P. Graceffa. Asbestos catalyzes hydroxyl and superoxide radical generation from hydrogen peroxide. *Arch. Biochem. Biophys.* **228,** 373–376 (1984).

121. E. D. Willis. Mechanisms of lipid peroxide formation is tissues. Role of metals and hematin proteins in the catalysis of the oxidation of unsaturated fatty acids. *Biochim. Biophys. Acta* **98,** 238–251 (1965).

122. G. W. Winston, D. E. Feierman, and A. I. Cederbaum. The role of iron chelates in hydroxyl radical production by rat liver microsomes, NADPH-cytochrome P450 reductase and xanthine oxidase. *Arch. Biochem. Biophys.* **232,** 378–390 (1984).

123. C. C. Winterbourn. Comparison of superoxide with other reducing agents in the biological production of hydroxyl radicals. *Biochem. J.* **182,** 625–628 (1979).

124. C. C. Winterbourn. Hydroxyl radical production in body fluids: Roles of metal ions, ascorbate and superoxide. *Biochem. J.* **198,** 125–131 (1981).

125. C. C. Winterbourn. Lactoferrin-catalyzed hydroxyl radical production: additional requirement for a chelating agent. *Biochem. J.* **210,** 15–19 (1983).

126. C. C. Winterbourn. Free radical production and oxidative reactions of hemoglobin. *Environ. Hlth. Perspect.* **64,** 321–330 (1985).

127. C. C. Winterbourn, U. Benatti, and A. De Flora. Contributions of superoxide, hydrogen peroxide, and transition metal ions to autoxidation of the favism-inducing compound, divicine, and its reactions with hemoglobin. *Biochem. Pharmacol.* **35,** 2009–2016 (1986).

128. C. C. Winterbourn, W. B. Cowden, and H. C. Sutton. Autooxidation of dialuric acid, divicine, and isouramil: Superoxide dependent and independent mechanisms. *Biochem. Pharmacol.* **38,** 611–618 (1989).

129. C. C. Winterbourn and R. Munday. Glutathione-mediated redox cycling of alloxan. Mechanisms of superoxide dismutase inhibition and of metal-catalyzed hydroxyl radical production. *Biochem. Pharmacol.* **38,** 271–277 (1989).

130. C. C. Winterbourn and H. C. Sutton. Hydroxyl radical production from hydrogen peroxide and enzymatically generated paraquat radicals: catalytic requirements and oxygen dependence. *Arch. Biochem. Biophys.* **235,** 116–126 (1984).

131. C. C. Winterbourn and H. C. Sutton. Iron and xanthine oxidase catalyze formation of an oxidant species distinguishable from OH˙: comparison with the Haber-Weiss reaction. *Arch. Biochem. Biophys.* **244,** 27–36 (1986).

132. S. P. Wolff and R. T. Dean. Glucose autoxidation and protein modification. The potential role of "autoxidative glycosylation" in diabetes. *Biochem. J.* **245,** 243–250 (1987).

From: *Trace Elements, Micronutrients, and Free Radicals* • Ed.: I. E. Dreosti • ©1991 The Humana Press Inc.

CHAPTER 4

Dietary Prooxidants

MARIO UMBERTO DIANZANI

ABSTRACT

The oxidative breakdown of membrane lipids is considered a possible mechanism of irreversible injury of the cell. The present review analyzes the actual knowledge so far achieved on different dietary forms of cell damage, giving special emphasis to the proofs in favor of the pathogenetic involvement of lipid peroxidation. The alimentary conditions characterized by increased lipid peroxidation levels are the supplementation with polyunsaturated fatty acids, the ethanol intoxication, the iron overload, and the feeding rich in orotic acid. Whereas a conclusive relationship between lipid peroxidation and cell damage has not yet been obtained in these experimental conditions, an increasing bulk of data strongly support such a correlation.

1. INTRODUCTION

The discovery that lipid peroxidation is increased in vitamin E-deficient animals afforded a new approach to the understanding of the mechanism of action of this vitamin. At the same time, it introduced the concept that cell damage may be provoked by increased lipid peroxidation. After this first discovery, the major input to the study of lipid peroxidation as a general mechanism for cell damage came from the demonstration that this type of derangement occurs in liver cells very early after giving carbon tetrachloride (CCl_4) to animals (21,35,114,115,

77

Table 1
Pathological Conditions Characterized
by Increased Lipid Peroxidation in Liver Cells

I. Generally accepted as pathogenetically relevant	Haloalkane and haloalkene treatments (CCl_4, $CBrCl_3$, $CHCl_3$, 1,2-dibromoethane, trieline after induction, halothane after induction, and so on)
	Paracetamol treatment
	Treatment with iron in vitro
	Treatments with adriamycin
	Treatment with ethoxycoumarin
	Treatment with allyl alcohol
	Treatments with H_2O_2 in vitro, or with systems generating oxygen radical species (xanthineoxidase, acetaldehyde oxidase)
	Treatments with ethanol (chronic; acute to a minor extent)
	Cumene hydroperoxide addition in vitro
	Bromobenzene at necrotizing concentrations
II. Under discussion as far as regards the pathogenetic relevance	Tert-butyl-hydroperoxide treatment in vitro
	Sodium vanadate treatment in vitro
	Bromobenzene at low concentrations
	Choline deficiency
	Feeding diets containing high PUFA concentrations
	Feeding diets containing 1-1.5% orotic acid
III. Considered devoid of pathogenetic relevance	Menadione treatment; treatments with paraquat or diquat; treatment in vitro with diethylmaleate or with iodoacetamide; treatments in vitro with Cu, Cd, or Hg

134,135). The poisons that, at present, are thought to increase lipid peroxidation are reported in Table 1.

A description of the different dietary forms of cell damage possibly related to high lipid peroxidation must necessarily move from the Hartroft and Porta's concept (*62*) of the peroxidative balance. According to this concept, an equilibrium normally exists in lipid peroxidation level, occurring in the different tissues. This equilibrium results from the ratio between saturated and unsaturated fatty acids contained in the given tissue, multiplied by the ratio between the level of the antioxidant vs prooxidant substances (Fig. 1). So, the rate of lipid peroxidation is stimulated by a high content of polyunsaturated fatty acids (PUFA), as well as by a relative increase in prooxidants, whereas it is lowered by a relative increase in saturated fatty acids or of antioxidants.

As most fatty acids contained in tissue lipids come from the diet, and this is also the major source of both antioxidants and prooxidants, it is clear that the state of the peroxidative balance largely depends on the changes in the diet itself.

$$\text{Lipid peroxidation} = \frac{\text{Polyunsaturated lipids}}{\text{Saturated lipids}} \times \frac{\text{Pro-oxidants}}{\text{Antioxidants}}$$

Peroxidative balance

Fig. 1. The concept of the peroxidative balance according to Porta and Hartroft (62).

2. LIPID PEROXIDATION IN RELATION TO INTAKE OF POLYUNSATURATED FATTY ACIDS (PUFA)

Presently on this subject, we have only limited data regarding the damage produced in the liver by experimental ingestion of diets very rich in PUFA.

First, we must distinguish among changes seen in the treated animals, those really depending on increased lipid peroxidation inside the cells, from those produced by the ingestion of products of lipid peroxidation taking place in the food before intestinal absorption. It is well known that fats or oils containing high amounts of PUFA easily undergo lipid peroxidation, and that the addition of antioxidants to such fats is a routine commercial practice. Several disturbances have been described so far after ingestion of previously peroxidized oils or of their peroxidation products, as hydroperoxyepoxides and lipoperoxides (65,68,69,161) and hydroperoxyalkenals (163). A description of enzymatic damage provoked by the aldehydes when added in vitro, or given intraperitoneally, is given in appropriate reviews (29–31). This type of damage, however, remains out of the scope of the present chapter.

A high intake of PUFA can be achieved either by feeding a diet containing a normal fat content with high PUFA, or by administrating a high fat diet containing a relatively normal PUFA percentage. In the last case, the high PUFA intake results from high fat intake. Rather few experimental data are available about the consequences of ingesting a diet containing a normal fat content with a high percentage of PUFA. The damage so far produced in liver cells is, in any case, not great, if at all. Recent experiments by Draper et al. (45) have shown that feeding highly unsaturated cod liver oil, containing high amounts of eicosapentaenoic and eicosahexaenoic acids for 48 h results in an elevated elimination of malonyldialdehyde (MDA) by the urine. This effect is prevented by administration of vitamin E or of other antioxidants, such as butylhydroxytoluene (BHT) or vitamin C.

This represents a clear proof of the existence of increased lipid peroxidation in the tissues of animals receiving cod liver oil, but evidence for the presence of cell damage is still unclear. The possible cell damage is expected to occur rather late after feeding on this diet, because of the presence in the tissues of natural antioxidants, that need to be consumed before the damage of this type can develop.

The possibility that long-term feeding on such diets provokes liver damage is sustained by recent reports (9,112). Rao (112) found a promoting activity on liver carcinogenesis after feeding rats for some weeks on a PUFA-rich diet. Rats fed on this diet developed a rather high number of hyperplastic, gamma-glutamyltranspeptidase positive foci. Such foci, however, rarely progressed to hepatomas without any other treatment. Rao et al. (113,153) think that the PUFA diet-induced promotion may be the consequence of increased lipid peroxidation generating free radicals, as well as other toxic compounds, as are the aldehydes. So, cell damage, and especially necrosis, would be the real promoter.

The information we have on the consequences of ingestion of high fat diets is greater than that on the high intake of PUFA-rich diets. It is clear, first of all, that a high intake of PUFA, either as such or contained in high fat diets, provokes an increase in the demand for the natural antioxidants, and especially vitamin E (25,45). In the absence of higher levels of vitamin E, high fat diets may develop a derangement in the peroxidative balance, and a relative avitaminosis E. Usually, high fat diets do not contain proportionally higher amounts of the vitamin, especially in the cases when high fat is from animal origin. Olive or seeds oils, on the contrary, usually contain sufficient amounts of vitamin E. These become exhausted after long-term conservation, but commercial preparations usually contain other antioxidants, especially butylhydroxytoluene (BHT), butylhydroxyanisol (BHA), or propyl gallate (PG).

Long-term administration of high-fat diets usually provokes fatty liver. The reason for this is that excess chylomicrons circulating after fat meals provide to the peripheral tissues and to the liver an amount of free fatty acids (FFA) higher than normal, in conditions where the activity of lipoprotein lipase is preserved (*see* Dianzani, 27,28 for reviews). Fatty liver develops when the net amount of FFA arriving from outside exceeds the oxidative and secretive capacities. Excess fat becomes, therefore, stored within liver cell in form of fat droplets.

The enlargement of the cell, very often referred to as "ballooning," provokes an increase in the reciprocal pressure the neighboring cells exert on one another. Consequently, a decrease in the blood arriving through the sinusoid, and at the same time an obliteration of absorbing surface of the cells, will develop. After a certain time, therefore, the protein mass of the cytoplasm decreases. One of the topics emerging from this pathogenetic description is whether increased lipid peroxidation can occur and play a role in the production of the final picture.

An increase in total fat of the cell surely provokes a net increase in PUFA, usually not compensated for by a corresponding increased supply in antioxidants, especially vitamin E, whose life-span inside the cells is expected to be longer than that of nonnatural antioxidants, such as BHT, BHA, or PG. In fact, the last substances are usually quickly oxidized and destroyed by the drug metabolizing chain of the smooth endoplasmic reticulum, whereas vitamin E can be regenerated by vitamin C (79,101).

Direct measurements of lipid peroxidation in fatty livers caused by the described mechanisms are still missing. Recently, however, Draper and coworkers (25,45) have produced evidence for increased urinary (MDA) elimination in rats receiving a high PUFA diet for a short time. Moreover, we get indirect evidence from the studies done on other types of fatty livers, and especially from those related to choline-deficiency. In fact, choline-deficient diets are usually also hyperlipidic, with normal vitamin E.

In the liver of rats force-fed a choline-deficient, high-fat diet, several years ago, Ugazio et al. (150) were able to find a net increase in the amount of fatty acids bound covalently to proteins (the so-called un-saponifiable fraction of fat). The fat extracted from liver homogenates displayed a rather high diene conjugation band, especially high at 168 h after force-feeding. These observations were considered as proof of increased lipid peroxidation in total fat, probably deriving from a change in the peroxidative balance. Binding of lipid free radicals with proteins to form insoluble complexes had been described by Roubal and Tappel (121,122) as the main mechanism for lipofuscin production in tissues. Lipofuscin is usually found in such types of fatty livers by histological methods. More recently, chemiluminescence methods have added further evidence about this point (48). In the experiments by Ugazio et al. (150), the diene conjugation band was not found to be increased in microsomal phospholipids. So, the conclusion was drawn that lipid peroxidation with consequent lipofuscin formation occurs mainly in the depot fat. Ugazio et al. (150) also concluded that fat accumulation in choline-deficiency does not depend on lipid peroxidation, whereas on the contrary, this is the consequence of the accumulation of fat.

Monserrat et al. (92) and Ghoshal et al. (51) found that lipid peroxidation is increased in kidney homogenates of choline-deficient weanling rats before the development of renal necrosis, which occurs after 5 d of deficiency. Even in this case, however, the diene conjugation band was absent in both mitochondrial and microsomal phospholipids. More recently, the topic has been investigated again by Perera et al. (104), who measured diene conjugation in microsomal phospholipids of the liver from rats fed choline-deficient, high-fat diets with different degrees of unsaturation in their fat. When the unsaturation level was high, the diene conjugation band was evident as early as 2 d after feeding on the diet, whereas it was lacking in microsomal phospholipids of rats receiving saturated fat. In this last case, however, the band appeared after one week feeding on the diet and remained always much lower than that seen after the same time in rats receiving the highly unsaturated fat.

Choline-deficiency acts as an initiator, and at the same time as a

promotor, in cancerogenetic models (*1,10,16,18,49,50,54,80,86,94,103,
104,112,125,130–132,145*), the major effect being reached when the given
fat is highly unsaturated. So, the hypothesis has been put forward that
the cancerogenetic effect in this model may be in some way related to
increased lipid peroxidation. It is noteworthy that in rats fed the choline-
deficient high-fat diet, increased lipid peroxidation is even shown by the
increased production of MDA by incubated homogenates (*53,132*).

There is, however, chronological disagreement between increased
urinary elimination of MDA and the peak in the appearance of the diene
conjugation band (*25*). Of course, this last result cannot be considered as
fully negative as far as it regards the main hypothesis, since we have no
idea about the dynamics of MDA urinary elimination.

Some doubts about the real meaning of increased lipid peroxidation
in choline deficiency arise, however, from recent results of Lombardi and
Banni (*86*), who found no diene conjugation band in microsomal phospholipids, and found on the contrary, high bands in total lipids extracted
from the whole liver, as well as from the diet itself. The opinion of the
authors is, therefore, that lipid peroxidation in liver fat is not the consequence of choline-deficiency itself, but only of the ingestion of preperox-
idized fat contained in the choline-deficient, high-fat diet. Further work
on this point is, therefore, necessary.

The problem is how to establish if this form of increased lipid
peroxidation can afford definite damage to liver cells. Of course, lipid
peroxidation occurring at membrane level might produce membrane
damage and, consequently, a change or loss of membrane function. Such
damage might even be the cause of cell death. In the case of fatty liver
resulting from high dietary fat intake, however, cell death is not the rule,
and full recovery of normal structures is achieved shortly after restora-
tion of normal diets. The choline-deficient, high fat diet, however, facilitates cell death provoked by chronic ethanol poisoning (*142,144*).

In the case of CCl_4-induced damage (*37,38,106–110*), as well as in
those produced by paracetamol (*3*) or by 1,2-dibromoethane (*4,146*),
acute cell death seems, however, to be a consequence of lipid peroxidation and especially of the action of the toxic aldehydes. In fact, cell death
is prevented by vitamin E enrichment of the hepatocytes, as well as by
addition of promethazine. Both promethazine and vitamin E are unable
to prevent covalent binding, but completely cancel lipid peroxidation and
the consequent aldehyde production (*33,36–39*), and at the same time,
acute cell death. Lack of cell death in fatty livers related to high-fat diets
may be only apparent, because of the rather long time needed for single
cells to die, as well as to the different general chronology of the experi-
ment. In any case, it has been shown that long-term choline-deficiency
provokes cell death (*1,53,54,90,94*) to an extent sufficient to enable Chandar et al. (*16*), as well as other authors (*54,162*), to think that cell death
and consequent regeneration are the real promoting mechanisms in
choline-deficiency. It is possible that this peculiar late cell death, con-

trasting with the acuteness of that occurring after CCl_4, results from the low rate of diffusion of toxic compounds (especially aldehydes) from peroxidizing lipids. In fact, in the case of CCl_4 these are mostly represented by microsomal phospholipids, freely contacting the soluble portion of the cytoplasm, whereas in the case of choline-deficiency, they are mostly represented by triglycerides (TG) contained in the large fat droplets, that are relatively segregated from the cytoplasm. One has to remember that the most toxic aldehydes, such as 4-hydroxy-nonenal and 4,5-dihydroxy-decenal, are rather lipophilic, because of the length of their chain, and are expected to diffuse from lipids less easily than the short-chained aldehydes.

Among the damage found in liver cells of rats fed a high-fat choline-deficient diet, Betschart and coworkers (*10,130*) have described consistent changes in surface receptors for insulin, glucagon, and epidermal growth factor. The authors advance the hypothesis that these changes are related to lipid peroxidation, and have, therefore, proposed that disregulation of cell growth and promotion may be concerned with such changes. If these results might be expanded to other receptors, and especially to those concerned with absorption of nutrients, this might well be a further reason for total decrease of cytoplasmic mass in long-term fatty livers. Of course, the data by Betschart and colleagues (*10,130*), refer to a condition in which both high fat intake and choline deficiency cooperate in the production of the damage.

The participation of liver peroxidation in the production of damage in fatty livers has recently been extended by Draper and coworkers (*25,45*) to include that following the administration of adrenocorticotrophic hormone, or of epinephrine. In these cases, fatty liver is the consequence of increased lipolysis in adipocytes. This depends on the activation of adipocytes lipase by increased cAMP levels. cAMP is synthesized by adenylate cyclase, an enzyme activated by adrenocorticotrophic hormone, cortisol, epinephrine, glucagon, and prostaglandins (*see* Dianzani, *28,29* for references). A strong release of ACTH from the hypophysis or of epinephrine from the adrenals, as well as of cortisone from the adrenal cortex, occurs during stress. So, clinical fatty livers following different forms of stress belong to this group, and result from increased FFA uptake from blood plasma. In these conditions, no apparent increased vitamin E uptake occurs, so increased fat may represent the reason for a change in the peroxidative balance. The experiments by Draper and associates (*25,45*) need further confirmation by methods other than elimination of urinary MDA. It has been recently shown that MDA and 4-HNE (4-hydroxynonenal) released from peroxidizing lipids become bound to serum proteins, and especially to albumin, where they can be detected and measured by spectrofluorimetric methods (*107*). The formation of the adduct between 4-HNE and albumin has been studied, even in vitro, and it has been demonstrated that the adduct is in equilibrium with the constituents (*23,24*). So, MDA and 4-HNE produced in a given organ during local damage can become trapped by serum proteins, transported by the blood stream and released in a distant area. In alcoholic humans, the amount of MDA and 4-HNE bound to serum albumin increases sharply (Poli et al., unpublished results). So, the comparison of values obtained by this method with those of the urinary elimination might be very important in order to understand both the presence of increased lipid peroxidation and its possible participation in tissue damage.

In conclusion, it seems very probable that increased lipid peroxidation occurs in all forms of fatty livers, characterized by a net increase of PUFA in the liver, on the condition that the increase in PUFA is not accompanied by a corresponding increase in vitamin E or other antioxidants.

There is, however, no definite evidence as yet that increased lipid peroxidation plays a role in the mechanism of cell damage in these conditions.

3. ETHANOL-INDUCED LIVER DAMAGE

It is well known that ethanol produces two types of liver damage, i.e., that following a single administration of a high dose of ethanol and that seen after chronic treatment. The acute ethanol-induced fatty liver is characterized by increase in liver TG, occurring 5–6 h after the administration of 6–8 g ethanol/kg body wt in the rat (*see* Dianzani (32), for references) and reaching a peak at about 12 h. After 24 h, normal TG levels are usually restored. Fatty liver is the main change observed in acute poisoning, and cell death is practically lacking.

Chronic treatment, mimicking the situation found in heavy drinkers, also produces a fatty liver, but this is complicated by the presence of cell death, as well as by other morphological changes, including the production of the alcoholic hyaline forming the Mallory bodies. Moreover, cell death is followed by regeneration, and very often, the picture moves towards alcoholic hepatitis, cirrhosis, and cancer. The suppression of ethanol intake does not always produce full recovery, the extent of this being entirely dependent on the stage of liver lesion. This has been studied especially in baboons, that are at present the unique animal model able to develop cirrhosis and cancer, like in humans (81–83). Several causes certainly interact to produce the picture of chronic ethanol induced liver damage. A decrease in the daily dietary intake of food, provoking protein and vitamin deficiencies (including vitamin E deficiency) occurs very often. It can play a role in decreasing apolipoprotein synthesis and, therefore, lipoprotein secretion. Moreover, phospholipid deficiency in the diet provokes choline deficiency. As choline is an important portion of lecithins, that are important constituents of membranes as well as of lipoproteins, this may represent a further reason for TG accumulation.

It has also been reported by Takeuchi's group (*142–144*), that the choline dietary requirement increases in chronically alcohol-treated rats. This may be the result of cell regeneration following hepatocyte death, or also to higher membrane destruction related to increased lipid peroxidation, or even to other unknown mechanisms. Other causes for TG increase are represented by the direct action of acetaldehyde or of other aldehydes on the microtubular system, or also by decreased production of ATP needed for protein synthesis, as well as for intracellular movements.

In any case, the bulk of the difference between acute and chronic ethanol-induced liver damage is represented by cell death, that is present only in the second case and seems, therefore, to be the key point for further progression of the damage. The speculative point, therefore, is why and how the same substance does not produce cell death when given in a single dose, whereas it does when given repeatedly. Several years ago, the hypothesis was advanced that increased lipid peroxidation, which occurs especially in chronic treatment, may have some bearing on the whole picture. The first demonstration of increased lipid peroxidation in the liver of animals treated acutely was an indirect one. In fact, Di Luzio and coworkers (*40–43*) were able to show that pretreatment with antioxidants decreased TG accumulation in the liver after acute ethanol. After the discovery that CCl_4-treatment produced an increase in lipid peroxidation both in vivo (*21,35*) and in vitro (*21,35,52,114,115,134*), as shown both by increased MDA production (*21,35*) and by the appearance of the diene conjugation band in microsomal phospholipids (*52,114,115*), Di Luzio moved to the direct measurement of the level of lipid peroxidation in liver homogenates and subcellular particles. Thus, it was possible (*20*) to show an increased production of MDA by liver homogenates both in the case of rats treated with ethanol in vivo, or of homogenates incubated in the presence of added ethanol. Hashimoto and Recknagel (*63*), however, were unable to find the diene conjugation band in microsomal phospholipids. Dianzani (*26*) and Torrielli et al. (*147*) were able to confirm both the increased production of MDA by homogenates, as shown by Comporti et al. (*20*), and the lack of the diene conjugation band in microsomal phospholipids. They explained the increase of the MDA production by a possible decrease in the liver concentration of natural antioxidants. In fact, they showed that the lag phase between the onset of MDA production by homogenates is much shorter in ethanol-treated than in normal homogenates.

In the meantime, several other authors reported increased MDA production after either acute (*19,44,70,74,87,88,142,154–157*), or chronic pretreatment (*75,77,116,117,127,128*). The results are especially convincing in the last case, in which the increase in lipid peroxidation is accompanied by a decrease in the intracellular concentration of the reduced glutathione (*85,138,149,154,155*). Measurements of ethane or pentane exhalation from ethanol-treated rats (*84*) or from alcoholic patients, (*105*), as well as from perfused livers (*93*), showed a net increase in comparison with non-treated specimens or with nonalcoholic patients. More recently, it has been found that MDA and 4-HNE, as well as other aldehydes coming from lipid peroxidation, accumulate in a higher amount after acute or chronic ethanol treatment (32,34). So, the positive nature of the TBA test is really concerned with increased lipid peroxidation. Moreover, MDA production by homogenates after ethanol increases in a parallel way with the decline in GSH. Practically all disappearing GSH is recovered in this system as G-S-S-G, so showing that acetaldehyde sequestration of GSH does not occur (120).

Acetaldehyde, however, stimulates MDA accumulation more than ethanol itself at a corresponding concentration (93,100,141).

Ethanol-induced stimulation of MDA production is inhibited, however, to a large extent by addition of catalase to the system. This suggests the involvement of hydrogen peroxide in the mechanism of this stimulation. Ethanol provokes an increase in MDA production, even if added in vitro to isolated hepatocytes. In the same system, acetaldehyde provokes again a more intense stimulation than ethanol (6,34).

At present, there is no doubt about the fact that ethanol and acetaldehyde stimulate lipid peroxidation. As far as it concerns ethanol, additional evidence comes from the demonstration of high chemiluminescence IN SITU in rat liver after acute administration (156), as well as by the studies regarding biliary release of glutathione disulphide (133). Esterbauer et al. (48) have recently pointed out the participation of 4-HNE in the formation of cellular chromolipids responsible for tissue chemiluminescence. Moreover, Corongiu et al., (22) have been able to find a diene conjugation band in microsomal phospholipids after acute ethanol by using a more sophisticated and careful method of estimation.

The remaining problems are whether ethanol and acetaldehyde stimulate lipid peroxidation, and to what extent this participates in the production of the damage. As far as it concerns the first problem, it is now clear that the increase in lipid peroxidation is a consequence of ethanol metabolism. Ethanol is oxidized in the liver to acetaldehyde by four main pathways.

1. By ethanol dehydrogenase, a cytosolic NAD-dependent enzyme. Oxidation by this pathway accounts for about 90% total oxidation of this substrate in normal conditions, i.e., in animals never treated with ethanol before. Acetaldehyde produced by the enzymatic action can be further oxidized by acetaldehyde dehydrogenases. These are flavoenzymes mostly sited in mitochondria. Peroxisomal xanthineoxidase, however, is also able to oxidize acetaldehyde. This can be reduced back to ethanol by ethanol dehydrogenase working in anaerobiosis. Other enzymes able to attack acetaldehyde are aldehyde dehydrogenases and glutathione transpeptidase. There are several isoforms of the former enzyme, requiring either NAD or NADP, as well as isoforms of al-

dehyde reductase, even in this case, either NAD- or NADP-dependent.

2. By the NADP-dependent microsomal ethanol oxidizing system (MEOS). This system transfers hydrogen from the substrate to NADP, and then from NADPH + H^+ to FAD. The hydrogen bonded to $FADH_2$ is then transferred to a peculiar form of cytochrome P450 (66,67,76). It is noteworthy that from the flavin, an electron can bypass to oxygen, so, forming the free radical superoxide anion O_2^-. This happens especially when the electron flow along the chain is quick enough to saturate the ability of cytochrome-p450 to release them to oxygen. So, a substantial defect in the available oxygen can favor the production of O_2^-. This free radical can then work as a hydrogen acceptor, so forming H_2O_2. This is then transformed into H_2O and O_2 in the presence of catalase, an enzyme originally sited in peroxisomes.

3. Destruction of hydrogen peroxide by this enzyme results in the formation of acetaldehyde from ethanol (pathway number 3).

4. In the presence of iron, hydrogen peroxide and the superoxide anion can interact, so producing the highly reactive OH· free radical. This may occur either by the Fenton's or by the Haber-Weiss's reaction:

$$\text{I.} \quad H_2O_2 + Fe^{2+} <========> OH^- + OH\cdot + Fe^{3+}$$

Fenton's reaction

$$\text{II.} \quad H_2O_2 + O_2^- + Fe^{3+} <========> OH^- + OH\cdot + Fe^{2+}$$

Haber-Weiss's reaction

The likelihood of these reactions in living tissues has been criticized recently (64).

It is clear that, from the four main pathways for ethanol oxidation to acetaldehyde, the last three are directly concerned with the production of free radicals, whereas the first one, accounting normally for about 90% ethanol oxidation, is essentially nonradical, even if some free radical production can result from the oxidation of acetaldehyde, i.e., the product of the reaction. The microsomal pathways are poorly efficient in normal animals, but become much more active after induction of the drug metabolizing system, either by ethanol itself or by phenobarbital, or also by other inducers of this type. So, the animals treated with ethanol by a single administration oxidized ethanol mostly by the ethanol dehydrogenase nonradical pathway. Contrariwise, the animals treated repeatedly with ethanol oxidized this substrate to a considerable extent also by other pathways, that generate free radicals. This may explain why the stimulation of lipid peroxidation is much higher in chronically treated animals than in those receiving a single dose for the first time.

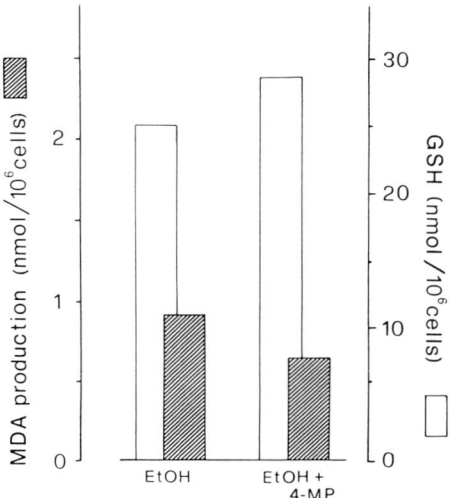

Fig. 2.   Inhibition by 4-methylpyrazol (4-MP) of the increase in production of malondialdehyde provoked by ethanol.

In the latter animals, the concurrent treatment with 4-methyl-pyrazol, an inhibitor of ethanoldehydrogenase, blocks to a great extent the increase in MDA production (7). This seems to show that in such rats, most stimulation of lipid peroxidation is concerned with that further oxidation of acetaldehyde, producing oxygen free radicals. 4-methyl-pyrazol is practically devoid of any effect when added to homogenates or microsomes, or even to hepatocytes, of rats previously treated repeatedly with ethanol (Fig. 2). This shows that in such animals, lipid peroxidation is stimulated by mechanisms other than acetaldehyde production and oxidation. We described above that in these animals, the radical micro-somal pathways for ethanol oxidation becomes quantitatively important. It is noteworthy that in deermice congenitally lacking ethanol dehy-drogenase, the effect of repeated doses of ethanol is not different from that observed in induced rats (83,129,143).

The possible production of an ethanol free radical during the metab-olism of this substrate by the microsomal pathways was first postulated by Slater (136). Several years ago, Cederbaum and coworkers (14,15,17) and Ingelman-Sundberg et al. (46,67) were able to show that during ethanol metabolism in microsomes, the very reactive OH˙ free radical is produced. The use of OH˙ free radical scavengers inhibits ethanol oxida-tion by the microsomal system, whereas the addition of iron is strongly stimulating. The authors suggested, therefore, that the OH˙ free radical produced by a Fenton or Haber-Weiss reaction in the presence of iron participates in the oxidation of ethanol in microsomes. The importance of iron in this system was further demonstrated by the discovery that ethanol oxidation is practically suppressed after chelation of iron by

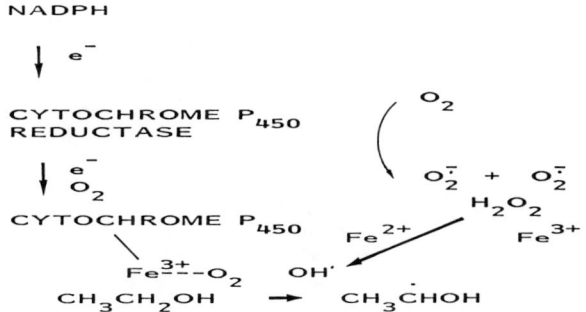

Fig. 3.   General scheme of the production of free radicals during ethanol oxidation in the smooth endoplasmic reticulum of liver cells.

desferroxiamine (desferal) (*13,14,77*). The importance of hydrogen peroxide in the mechanism results from the fact that sodium azide, which practically abolishes catalase activity, is strongly stimulatory of oxidation (*17,77*).

The authors, therefore, put forward the hypothesis that an ethanol free radical species may be formed during ethanol oxidation by microsomes. Albano et al. (*5–7*) succeeded in trapping this ethanol free radical by using 4-pyridyl-l-oxide-*t*-butyl nitrone as a spin trap. This radical was identified in microsomes from the liver of rats incubated in the presence of ethanol. The trapped signal surely belonged to ethanol, as the typical deformation of it introduced by the use of ^{13}C-containing ethanol overlapped with the signal obtained with ^{12}C-containing ethanol. It is now clear that the unpaired electron is sited in the carbon in number 1 position, and that the structure of a radical is that of a hydroxyethyl free radical. Its production is much higher when microsomes are isolated from rats receiving repeated doses of ethanol, or also of phenobarbital.

The signal is, however, abolished in the presence of desferal, and decreases strongly in the presence of the OH⋅ free radical scavengers, mannitol or benzoate. The addition of azide strongly intensifies the production of the signal, whereas the addition of catalase is strongly inhibitory. So, there is now clear evidence about the formation of the hydroxyethyl free radical during ethanol oxidation in vitro by microsomes in the presence of iron (Fig. 3). The problem whether the production of the free radical is entirely dependent on the interaction of ethanol with OH⋅ formed in the presence of iron, or if a certain amount of hydroxyethyl free radical can be produced directly during the electron flow along the microsomal chain, independently of OH⋅ formation, is still open (*7,126*). Albano et al. (*7*) never succeeded in completely abolishing the signal formation in the presence of excess desferal, so they conclude that at least 15% of the total hydroxyethyl free radical is produced in the chain by an OH⋅-independent mechanism.

The problem remains open about the real participation of this new free radical in the production of cell damage. We do not know at present the half-life of the new free radical, but it is expected to be longer than that of OH˙. So, from this point of view, ethanol can act as a scavenger of the dangerous and reactive OH˙ free radicals, independently of their mechanism of formation. In rats receiving a single and moderate dose of ethanol, this may scavenge OH˙ formed during the subsequent acetaldehyde oxidation, as well as in other reactions. However, in rats treated repeatedly, the production of reactive oxygen species becomes high, and the hydroxyethyl free radical may be produced in great amounts. So, there is a possibility of its participation in the damage.

Another point is to establish which type of damage free radicals can provoke in cells. A first type of damage comes from the stimulation of lipid peroxidation, probably started by the oxygen free radical species, and possibly by hydroxyethyl free radical. In fact, the reactions occurring in the smooth endoplasmic reticulum chain take place in a medium very rich in lipids, so the possibility for a quick interaction of OH˙ or O_2^- with local PUFA is consistent. One has to remember also that the half-life of O_2^- is shorter in a lipid than in an aqueous phase. Even the hydroxyethyl free radical is expected to display a higher solubility in lipids than in water.

The real meaning of lipid peroxidation in the pathological picture induced by ethanol is, however, still not well established. Probably, this process plays a minor role in the acute damage, essentially consisting of fat accumulation without necrosis. It is probable, however, that lipid peroxidation plays a role in the mechanism of the chronic lesions, where cell death is consistent. Lipid peroxidation certainly plays a role in the acute cell death as provoked by CCl_4, paracetamol, or 1,2-dibromoethane (3,4,106,107,110). Moreover, lipid peroxidation products, especially aldehydes, may play a role in producing damage both inside the cells where they are produced, or in the surrounding tissue. In fact, they have been proven to be able to escape in the surrounding medium from still living hepatocytes (106,107,110) and to accumulate in the tissues (151). As 4-HNE and other aldehydes of the toxic 4-hydroxy-2,3-trans unsaturated series formed during lipid peroxidation (47,110), display chemotactic activity toward polymorphonuclear leukocytes (23,24), it seems possible that they participate in recruitment of inflammatory cells.

Moreover, they can crossreact with components of the connective tissue. The increased production of aldehydes in alcoholic humans has been clearly demonstrated by Poli et al. (paper submitted for publication). So, we can now accept the idea of the possible participation of lipid peroxidation in the mechanism of the chronic liver lesions. However, we have no idea at present concerning the extent and the weight of this participation.

4. NUTRITIONAL IRON OVERLOAD

Iron is a trace element provided with pro-oxidant activity. As such, it may collaborate in producing cell damage when associated with other toxic substances (55). For instance, liver hemosiderosis has been described after ethionine administration, without any need for extra addition of excess iron to the food (71). Moreover, MacDonald (89) was able to produce liver hemosiderosis by giving a choline-deficient diet enriched with iron to rats. Richter (118,119) described massive liver siderosis after feeding a diet containing 1.23% ferric ammonium citrate. Rats were subjected to cyclic starvation feeding (3 d of fasting, followed by 3 d of feeding with iron-enriched diet) over periods ranging up to 6 mo. Other authors administered ferrocarbonyl (2.5% iron in the diet) (8). The effect of extra iron as added to the incubation medium of isolated hepatocytes has also been studied (39,109,110). The main purpose of all such investigations was to study if a large excess of the pro-oxidant metal would be able to increase lipid peroxidation in tissues, as well as to produce cell damage by itself, or by potentiation of that provoked by other substances as, for instance, ethanol or carbon tetrachloride. At the same time, it was found interesting to compare the lesions found in these conditions with those occurring in "spontaneous" hemochromatosis.

Richter (118,119) did not find any severe cell damage in cells after feeding rats on the 1.2% iron diet. The described damage was restricted to the accumulation of ferritin inside both macrophages and hepatocytes. Ferritin was mostly segregated within secondary lysosomes, but was evident even in the cytoplasmic matrix, especially near to the spaces of Disse, and in the sinusoid lumen. Richter thinks that some myelinic figures found in the Disse's spaces, as well as in the sinusoid lumens, originated from the aggregation of microvilli. The responsible process is not discussed, but the possibility exists that it is lipid peroxidation of cell membranes. In fact, Slivka et al. (137) reported the generation of OH' free radicals after receiving iron in the diet. In any case, Richter found no evidence for damage in the cytoplasm surrounding the ferritin masses. In the Bacon et al. model (8), iron overload produced siderosis of the liver, as well as other organs.

However, blood transaminases were not modified in any step of the treatment, therefore showing lack of necrosis. Biochemistry of liver cell did not show any important change either in mitochondrial or in lysosomal function. Even the enzymes of the smooth endoplasmic reticulum were unmodified. Lipid peroxidation was increased in terms of MDA produced by homogenates, but was unmodified in isolated microsomes. So, it seems clear that iron overload is not an important cause for cell damage. The fact that lipid peroxidation does increase in homogenates may be the consequence of the homogenization itself, producing deeper and more effective contact between iron and the lipoperoxidative system. In any case, iron is present in the organs of treated animals in the form of ferritin, where the element is in the trivalent state and, therefore, unable to stimulate lipid peroxidation without reduction.

The problem whether ferritin can release iron susceptible to be reduced and thereby able to stimulate lipid peroxidation has been discussed by Halliwell and coworkers (60,61) and by Gutteridge et al. (56–58). These authors found that in certain circumstances, divalent iron is released from ferritin. However, it is questionable whether such circumstances can occur in the whole cell. Recently, Dhanakoti and Draper (25) found increased urinary elimination of MDA in rats treated with iron nitrilotriacetate, so showing that iron produces somewhere (but not necessarily in the liver) an increase in lipid peroxidation. This model, however, needs deeper consideration and expansion.

Incubation of hepatocytes in the presence of either 100 μM (final concentration) ADP/FeCl$_3$ or of 100 μM FeSO$_4$, produces the rapid appearance in the cells of electron spin resonance signals, that can be easily trapped by POBN (108,109). This shows that free radicals are produced under these conditions. The analysis of the superfine splitting constants of the observed ESR signals, as compared with the signal produced after treatment with the soybean lipoxygenase of microsomal lipids, allowed their identification with lipodienyl-type radicals. So, the incubation of the isolated hepatocytes with either trivalent or divalent iron produces a big increase in lipid peroxidation. The production of the signal is prevented if the hepatocytes were isolated from rats receiving high amounts of vitamin E. The kinetics of lipid peroxidation occurring in hepatocytes in the presence of ADP/iron has been studied in depth, and found to be somewhat different from that induced in hepatocytes by CCl$_4$ (48,106) (Fig. 4). In fact, not only the total amount of carbonyls so far produced was much higher in the presence of iron than in that of CCl$_4$, but in the first case, there was a relatively higher production of MDA than of other aldehydes, including the very toxic 4-hydroxy-2,3-trans-unsaturated series. Nevertheless, the total amount of the last compounds exceeded by far that produced in the presence of CCl$_4$. Moreover, in the case of CCl$_4$, there was a certain amount of PUFA that had disappeared and that could not be recovered in the form of carbonyls.

Irreversible damage of the hepatocytes, as monitored in terms of the dye exclusion test, or also by release in the medium of lactate dehydrogenase (LDH), started after 2 h incubation, both in the presence of CCl$_4$ or with ADP-iron. At this time, it was higher in the case of cells treated with iron than in that of the cells treated with CCl$_4$. At 3 and 4 h, however, it became much more evident in the latter case, independently of the fact that aldehyde production was much higher in the presence of iron than in the presence of CCl$_4$. When the hepatocytes were isolated from rats previously given vitamin E (100 mg/kg body wt ip, 15 h before hepatocyte isolation), the release of LDH was very low, both in the case

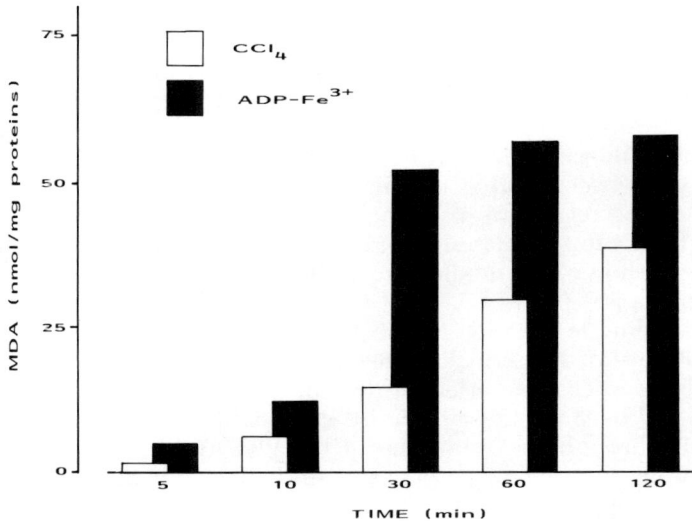

Fig. 4.   Difference in the kinetics of lipid peroxidation as stimulated either by CCl_4 or by ADP-iron in isolated hepatocytes.

of CCl_4 and of ADP-iron. This shown that lipid peroxidation plays a large role in the mechanism of this type of cell death, but at the same time, demonstrates that CCl_4, which is less pro-oxidant than ADP-iron, may even act by an additional mechanism(s). This might well be covalent binding of carbon tetrachloride-derived free radicals (haloalkylation), but the possibility of a different mechanism cannot be set aside. As far as other metabolic derangements are concerned, ADP-iron or Fe^{2+} produce a decrease in glucose-6-phosphatase activity in hepatocytes, as well as a derangement in protein synthesis and in lipoprotein secretion. Mitochondria are severely damaged too (8).

These changes are much less consistent than those seen (109) in the presence of CCl_4, and occur later. Moreover, ADP-iron, as different from CCl_4 does not provoke the early discharge of calcium from mitochondria as does CCl_4 treatment (2,38,39,107). In mouse liver, at micromolar levels, iron provokes the activation of a membrane-bound protein kinase (85). In conclusion, these experiments show that lipid peroxidation produces a lot of cell damage in these cell systems, but that seen after CCl_4 is not uniquely related to lipid peroxidation.

The fact that oral iron overload does not produce any serious damage to the liver, whereas the incubation of isolated hepatocytes with the metal produces severe damage and cell death, is at present difficult to explain. The only possible approach seems to be that iron reaches the liver cells, in the intact animal, in a form unable to produce damage. It is well known that circulating iron is bound to transferrin. So, it seems possible that in this form, iron cannot stimulate lipid peroxidation. The

problem arises, therefore, whether free iron can be released from transferrin or ferritin (at least in some circumstances), both outside or inside
the cells, in order to be able to increase lipid peroxidation.

According to Saito et al. (124), transferrin could release iron to ADP
at low pH values; in the presence of xanthineoxidase as a reducing agent,
iron would become able to stimulate lipid peroxidation. At pH 7.0,
however, the release of iron from transferrin is minimal, and probably
not sufficient to stimulate lipid peroxidation. The release of iron would
be possible however, from ferritin in the presence of xanthineoxidase (11)
at physiological pH.

According to Farber (140) the release of iron from ferritin is inhibited
by inhibitors of lysosomal proteases. This would show that it occurs
inside lysosomes as an effect of protein catabolism. Iron released in this
way would become responsible for cell death in the presence of H_2O_2,
probably through the formation of the OH˙ free radical.

Young et al. (164), however, found that very little ^{59}Fe bound to
transferrin can be recovered inside lysosomes after incubation of the
complex with isolated hepatocytes. After 2 h incubation, about 70% iron
is found associated with ferritin; the residual portion is thought by the
authors to represent an intracellular transit pool. The uptake of ^{59}Fe
bound to transferrin decreases in hepatocytes in the presence of high (10
m*M*) concentration of ethanol (99).

The involvement of excess dietary iron with cell damage has been
proposed in the case of ethanol, and indeed, Valenzuela et al. (152) and
Nordmann et al. (95) reported heavier liver damage in rats treated at the
same time with iron and ethanol. Mazzanti et al. (91) have proposed an
increased intestinal absorption of iron as a mechanism of liver damage in
ethylists. In our hands (7,38) the addition of iron to microsomes was able
to strongly intensify the ESR signal of hydroxyethyl free radicals. Moreover, by using single hepatocytes, the presence of iron in the incubation
medium is required in order to trap free radical signals. Their nature,
however, has not yet been established. So, although there is some
evidence of an interaction between iron and ethanol in the production of
cell damage, the proof is still inconclusive.

5. DIETS RICH IN OROTIC ACID

Several years ago, Standerfer and Handler (139) described a fatty
liver following administration to rats for a few days of a diet containing
1% orotic acid, a pyrimidine. Novikoff et al. (96) studied the pathogenesis of this process and found that fatty liver was preceded by disor-
ganization of the endoplasmic reticulum. Subsequent work by the
Novikoff's group (97,98), as well as by other authors, (111,123) showed
that this type of fatty liver was mostly related to a decline in lipopro
tein secretion. As total protein synthesis is increased, the hypothesis was

put forward that the block in lipoprotein secretion was related to a decline in the synthesis of a specific apolipoprotein (*158,160*). Triglyceride accumulation, as well as the block in apolipoprotein synthesis, are prevented by the administration of adenine (*159*). So, it is generally thought that this type of fatty liver is the consequence of an inbalance between purines and pyrimidines, favoring the synthesis of certain mRNAs at the expense of others. This interpretation, however, is not entirely convincing, since excess of other types of pyrimidines or purines does not produce TG accumulation.

Kinsella (*72,73*) studied lipid peroxidation in the livers of rats receiving for a week a diet rich in orotic acid, and found increased production of MDA by homogenates. He attributed the stimulation of lipid peroxidation to the high oxido-reductive potential of the couple, orotic–dihydroorotic acid. Torrielli and coworkers (*147,149*) confirmed the increased production of MDA, but were unable to reveal the appearance in microsomal phospholipids of the diene conjugation band showing the preexistence of lipid peroxidation in vivo. In their opinion, the higher MDA production might be related to a decrease in the endogenous antioxidant levels provoked by orotic acid.

More recently, Haines and Tokmakjian (*59*) suggested a mechanism of action for orotic acid based on a change in the regulation of the synthesis of phosphatidylcholine. In fact, rats receiving this pyrimidine for one day show a great increase in the level of cytosine triphosphate in liver cells. It has also been found that feeding a diet containing 1% orotic acid for five weeks is a good promoter of carcinogenesis in rat liver models (*78*). Alkaline elution and alkaline sucrose gradients, as well as viscosimetric studies, have suggested that this phenomenon is mostly concerned with changes in the DNA conformation, more than in DNA strand breaks (*102,113,153*). The reasons for these changes are still unknown, and the possibility, therefore, exists of free radical production during the metabolism of orotic acid. This matter is, however at present, completely speculative.

REFERENCES

1. S. E. Abanobi, B. Lombardi, and H. Shinozuka. Stimulation of DNA synthesis and cell proliferation in the liver of rats fed a choline-devoid diet and their suppression by phenobarbital. *Cancer Res.* **42**, 412–415 (1982).
2. E. Albano, G. Bellomo, R. Carini, F. Biasi, G. Poli, and M. U. Dianzani. Mechanisms responsible for carbon tetrachloride-induced perturbation of mitochondrial calcium homeostasis. *FEBS* **192**, 184–188 (1985).
3. E. Albano, G. Poli, E. Chiarpotto, F. Biasi, and M. U. Dianzani. Paracethamol-stimulated lipid peroxidation in isolated rat and mouse hepatocytes. *Chem.–Biol. Interactions* **47**, 249–263 (1983).
4. E. Albano, G. Poli, A. Tomasi, A. Bini, V. Vannini, and M. U. Dianzani. Toxicity of 1,2-dibromoethane in isolated hepatocytes: role of lipid peroxidation. *Chem.–Biol. Interactions* **50**, 255–265 (1984).

5. E. Albano, A. Tomasi, L. Goria-Gatti, R. Carini, V. Vannini, and M. U. Dianzani. Free radical metabolism of ethanol, in *Free Radicals, Cell Damage and Disease,* C. Rice-Evans, ed., Richelieu Press, London, 1986, pp. 117–126.

6. E. Albano, G. Poli, A. Tomasi, L. Goria-Gatti, and M. U. Dianzani. Free radical formation and oxidative stress during hepatic ethanol-intoxication, in *Pathophysiology of the liver* (M. U. Dianzani and P. Gentilini, eds.), Excerpta Medica, Amsterdam, 1988, pp. 3–13.

7. E. Albano, A. Tomasi, L. Goria-Gatti, G. Poli, V. Vannini, and M. U. Dianzani. Free radicals metabolism of Alcohols by rat liver microsomes. *Free Rad. Res. Comms.,* **3,** 243–249, 1987.

8. B. R. Bacon, R. O'Neill, and C. H. Parker. Iron-induced peroxidative injury to isolated rat hepatic mitochondria. *J. Free Radicals in Biology and Medicine* **2,** 339–347 (1986).

9. H. Bartsh, M. Ahotupa, A. M. Camus, J. C. Berezat, and E. Hietanen. Modulation of nitrosamine-induced cancers in rats by polyunsaturated lipid diets; effects of prooxidant state and lung metabolism, in *Models and Mechanisms in Chemical Carcinogenesis,* IV Sardinian Meeting, Alghero, October 23–27, 1987, 128A.

10. J. M. Betshart, M. A. Viryi, M. I. R. Perera, and H. Shinozuka. Alteration in hepatocyte insulin receptors in rats fed a choline-deficient diet. *Cancer Res.* **46,** 4425–4430 (1986).

11. B. J. Bolann and R. J. Ulvik. Release of iron from ferritin by xanthineoxidase. *Biochem. J.* **243,** 53–59 (1987).

12. J. M. Braugler, L. A. Duncan, and R. L. Chase. The involvement of Iron in lipid peroxidation. Importance of ferric to ferrous ratios in initiation. *J. Biol. Chem.* **261,** 10282–10289 (1986).

13. A. I. Cederbaum and G. Cohen, Inhibition of the microsomal oxidation of ethanol and 1-butanol by the free-radical spin-trapping agent 5,5-dimethyl-1-pyrroline-1-oxide. *Arch. Biochem. Biophys.* **204,** 397–403 (1980).

14. A. I. Cederbaum and E. Dicker, Inhibition of microsomal oxidation of alcohols and hydroxyl-radical-scavenging agents by the iron-chelating agent desferroxiamine. *Biochem. J.* **210,** 107–113 (1983).

15. A. I. Cederbaum, G. Miwa, G. Cohen, and A. Y. H. Lee. Production of hydroxyl radicals and their role in the oxidation of ethanol by a reconstituted microsomal system containing cytochrome-p_{450} purified from phenobarbital-treated rats. *Biochem. Biophys. Res. Comm.* **91,** 747–754 (1979).

16. N. Chandar, J. Amenta, J. C. Kandala, and B. Lombardi, Liver cell turnover in rats fed a choline-devoid diet. *Carcinogenesis* (London), **8,** 669–673 (1987).

17. G. Cohen and A. I. Cederbaum. Microsomal metabolism of hydroxyl-radical-scavenging agents: relationship to the microsomal oxidation of alcohols. *Arch. Biochem. Biophys.* **199,** 438–447 (1980).

18. A. Columbano, S. Rajalakshmi, and D. S. R. Sarma, Requirements of cell proliferation for the initiation of liver carcinogenesis as assayed by three different procedures. *Cancer Res.* **41,** 2079–2083 (1981).

19. M. Comporti, M. Benedetti, and E. Chieli, Studies on in vitro peroxidation of liver lipids in ethanol-treated rats. *Lipids* **8,** 498–502 (1973).

20. M. Comporti, A. Hartman, and N. R. Di Luzio. Effect on in vivo and in vitro ethanol administration in liver lipid peroxidation. *Lab. Invest.* **16,** 616–624 (1967).

21. M. Comporti, C. Saccocci, and M. U. Dianzani, Effect of CCl_4 in vitro and in vivo on lipid peroxidation of rat liver homogenates and subcellular fractions. *Enzymologia* **29,** 185–203 (1965).

22. F. P. Corongiu, M. Lai, and A. Milia. Carbon-tetrachloride, bromotrichloromethane, and ethanol acute intoxication. *Biochem. J.* **212**, 625–631 (1983).
23. M. Curzio, C. Di Mauro, H. Esterbauer, and M. U. Dianzani. Chemotactic activity of aldehydes. Structural requirements. Role in inflammatory process. *Biomedicine and Pharmacotherapy* **41**, 304–314 (1987).
24. M. Curzio, H. Esterbauer, C. Di Mauro, G. Cecchini, and M. U. Dianzani. Chemotactic activity of the lipid peroxidation product 4-hydroxynonenal and homologous aldehydes. *Biol. Chem. Hoppe Seyler* **367**, 321–329 (1986).
25. S. N. Dhanakoti and H. H. Draper. Response of urinary malondialdehyde to factors that stimulate lipid peroxidation in vivo. *Lipids* **22**, 643–646 (1987).
26. M. U. Dianzani, Azione patogena dell'alcool sulle cellule: la steatosi epatica da etanolo. *Metabolismo* **5**, 339–361 (1969).
27. M. U. Dianzani. Toxic liver injury by protein synthesis inhibitors in *Progress in Liver Disease* (H. Popper and F. Schaffner, eds.) Grune and Stratton, New York, 1975, **5**, pp. 232–245.
28. M. U. Dianzani. Biochemical aspects of fatty liver, in *Biochemical Mechanisms of Liver Injury* (T. F. Slater, ed.), Academic, New York, 1978, pp. 45–95.
29. M. U. Dianzani. Reactions of the liver to injury: fatty liver, in *Toxic Injury of the Liver* (E. Farber and M. Fisher eds.), Marcel Decker, New York, 1979, pp. 281–331.
30. M. U. Dianzani. Biological activity of methyl glyoxal and related aldehydes, in *Submolecular pathology and cancer*. Ciba Found. Symp. 67, Elsevier, Amsterdam, 1979, pp. 245–265.
31. M. U. Dianzani. Biochemical effects of saturated and unsaturated aldehydes, in *Free Radicals Lipid Peroxidation and Cancer*, (D. C. H. McBrien and T. F. Slater, eds.), Academic, London, 1982, pp. 129–158.
32. M. U. Dianzani. Lipoperoxidation in ethanol poisoning: a critical reconsideration. *Alcohol and alcoholism* **20**, 161–173 (1985).
33. M. U. Dianzani. The role of free radicals in liver damage. *Proc. Nutr. Soc.* **46**, 43–52 (1987).
34. M. U. Dianzani. Role of Free Radical-mediated Reactions in Ethanol-induced Liver Damage, in Alcohol Toxicity and Free Radical Mechanisms, ed. by R. Nordman, C. Ribière and H. Rouach, Pergamon Press, Oxford-New York, 1988, pages 35–41.
35. M. U. Dianzani, F. M. Baccino, and M. Comporti. The direct effect of carbon tetrachloride in subcellular particles. *Lab. Invest.* **15**, 149–156 (1966).
36. M. U. Dianzani, E. Chiarpotto, F. Biasi, and G. Poli. CCl_4-induced increase of hepatocyte free arachidonate level, in *Eicosanoids, Lipid Peroxidation, and Cancer* (S. Nigam and T. F. Slater, eds.), Springle Verlag, Berlin, 1988, p. 235–241.
37. M. U. Dianzani and G. Poli. Lipid peroxidation and haloalkylation in CCl_4-induced liver injury, in *Free Radicals and Liver Injury* (G. Poli, K. H. Cheeseman, M. U. Dianzani, and T. F. Slater, eds.), IRL Press, Oxford, 1985, pp. 149–157.
38. M. U. Dianzani and G. Poli. Lipid Peroxidation and Liver Cell Death, Intern. Meeting on Diet, Free Radicals, and Tissue Damage, November 19–20, Istituto Nazionale della Nutrizione, Roma, 1987, p. 20A.
39. M. U. Dianzani, G. Poli, and E. Albano. Sovraccarico di Ferro ed etanolo a livello di epatociti isolati. *Quaderni della Vite ed Enologia, Univ. Torino* **11**, 1–6 (1987).
40. N. R. Di Luzio. Prevention of the acute ethanol-induced fatty liver by the simultaneous administration of antioxidants. *Life Sci.* **3**, 113–120 (1964).

41. N. R. Di Luzio. The influence of intravenously administered hexahydro-coenzyme Q_4 on liver injury. *Life Sci.* **5**, 1467–1471 (1966).

42. N. R. Di Luzio. A mechanism of the acute ethanol-induced fatty liver and the modifications of liver injury by antioxidants. *Lab. Invest.* **15**, 50–63 (1966).

43. N. R. Di Luzio and F. Costales. Inhibition of the ethanol and carbon tetrachloride-induced fatty liver by antioxidants. *Exp. Mol. Pathol.* **4**, 141–154 (1964).

44. N. R. Di Luzio and T. E. Stege. The role of ethanol metabolites in hepatic lipid peroxidation, in *Alcohol and the Liver* (M. M. Fisher and J. G. Rankin, eds.), Plenum, New York, 1977, pp. 45–62.

45. H. H. Draper, L. Polensek, M. Hadley, and C. G. McGirr. Urinary malon-dialdehyde as an indicator of lipid peroxidation in the diet and in the tissue. *Lipids* **19**, 836–843 (1984).

46. G. Ekstrom, T. Cronholm, and M. Ingelman-Sundberg. Hydroxyl-radical production and ethanol oxidation by liver microsomes isolated from ethanol-treated rats. *Biochem. J.* **233**, 755–761 (1986).

47. H. Esterbauer, K. H. Cheeseman, M. U. Dianzani, G. Poli, and T. F. Slater. Separation and characterization in the aldehydic products of lipid peroxidation stimulated by ADP-Fe^{2+} in rat liver microsomes. *Biochem. J.* **208**, 129–140 (1982).

48. H. Esterbauer, E. Koller, R. G. Slee, and J. F. Koster. Possible involvement of the lipid peroxidation product 4-hydroxynonenal in the formation of the fluorescent chromolipids. *Biochem. J.* **239**, 405–409 (1986).

49. A. K. Ghoshal, M. Ahluwalia, and E. Farber. Cell death in rats fed a choline-deficient methionine-low diet. *Amer. J. Pathol.* **113**, 309–314 (1983).

50. A. K. Ghoshal and E. Farber. The induction of liver cancer by dietary deficiency of choline and methionine without added carcinogen. *Carcinogenesis* **5**, 1367–1370 (1984).

51. A. K. Ghoshal, A. J. Montserrat, E. A. Porta, and W. S. Hartroft. Role of lipoperoxidation in early choline deficiency. *Exp. Mol. Pathol.* **17**, 31–35 (1970).

52. A. K. Ghoshal and R. O. Recknagel. Positive evidence of acceleration of lipoperoxidation in rat liver by carbon tetrachloride: in vitro experiments. *Life Sci.* **4**, 1521–1530 (1965).

53. A. K. Ghoshal, T. H. Rushmore, and E. Farber. Initiation of carcinogenesis by a dietary deficiency of choline in the absence of added carcinogens. *Cancer Letters* **36**, 289–296 (1987).

54. L. H. Giambarresi, S. L. Katyal, and B. Lombardi. Promotion of liver cancer in the rat by a choline devoid diet: role of cell liver necrosis and regeneration. *Br. J. Cancer* **45**, 825–829 (1982).

55. M. H. N. Golden, B. E. Golden, and F. I. Bennet. Relationships of trace elements deficiencies to malnutrition, in *Nutrition and Children* (R. K. Chandra, ed.), Nestle' Nutrition, Vevey, Raven, New York, 1985, pp. 185–207.

56. M. C. Gutteridge. Ferrous iron-EDTA-stimulated phospholipid peroxidation. A reaction changing from alcoxy-radical to hydroxyl-radical-dependent initiation. *Biochem. J.* **224**, 697–701 (1984).

57. M. C. Gutteridge, B. Halliwell, A. Treffry, P. M. Harrison, and D. Blake. Effect on ferritin-containing fractions with different iron loading on lipid peroxidation. *Biochem. J.* **209**, 557–560 (1983).

58. M. C. Gutteridge, D. A. Rowley, and B. Halliwell. Superoxide-dependent formation of the hydroxyl-radicals and lipid peroxidation in the presence of iron salts. *Biochem. J.* **206**, 605–609 (1982).

59. D. S. M Haines and S. D. Tokmakjian. Actions of dietary orotic acid on liver synthesis of phosphatidylcholine and phosphatidylethanolamine. *Biochem. Cell Biol.* (Canada) **105**, 105–111 (1987).

60. B. Halliwell. Use of desferroxiamine as a probe for iron-dependent formation of hydroxyl-radicals. Evidence for a direct reaction between desferal and the superoxide radical. *Biochem. Pharmacol.* **34**, 229–235 (1985).

61. B. Halliwell and M. C. Gutteridge. Oxygen toxicity, oxygen radicals, transition metals, and disease. *Biochem. J.* **219**, 1–14 (1984).

62. W. S. Hartroft and E. A. Porta. Present knowledge of ceroid pigment, in *Present Knowledge in Nutrition*, 3rd Ed., The Nutrition Foundation, Washington D.C. 1967, pp. 28–31.

63. S. Hashimoto and R. O. Recknagel. No chemical evidence of hepatic lipid peroxidation in acute ethanol toxicity. *Exp. Mol. Pathol.* **8**, 225–242 (1968).

64. P. Hochstein. The formation of free radicals from quinoid compounds, Intern. Meeting on Diet, Free Radicals, and Tissue Damage, Istituto Nazionale della Nutrizione, Rome, November 19–20, 1987, Abstract 2.

65. T. Imagawa, S. Kasai, K. Matsui, and T. Nakamura. Detrimental effect of methyl hydroperoxy-epoxy-octadecenoate on mitochondrial respiration: detoxication by rat mitochondria. *J. Biochem.* **94**, 87–96 (1983).

66. M. Ingelman-Sundberg, G. Eckstrom, N. Tindberg, and I. Johansson. Lipid peroxidation is dependent on ethanol-inducible cytochrome-P-450 from rat liver. First Congress of the European Society for Biomedical Research on Alcoholism, Paris, September 18–19, 1987, Abstract 7.

67. M. Ingelman-Sundberg and I. Johansson. Mechanisms of hydroxyl radical formation and ethanol oxidation by ethanol-inducible and other forms of rabbit liver microsomal cytochromes-P-450. *J. Biol. Chem.* **259**, 6447–6458 (1984).

68. T. Kaneda and S. Ishii. Studies on the nutritive values of lipids. VIII. Nutritive value or toxicity of highly unsaturated fatty acids. *Bull. Japan. Soc. Sci. Fish* **19**, 171–177 (1953).

69. T. Kaneko, S. Honda, S. I. Nakano, and M. Matsuo. Lethal effects of a linoleic acid hydroperoxide and its autooxidation products, unsaturated aliphatic aldehydes, on human diploid fibroblasts. *Chem.–Biol. Interactions* **63**, 127–137 (1987).

70. Y. Kera, S. Komura, Y. Ohbora, T. Kiryama, and K. Inoue. Ethanol-induced changes in lipid peroxidation and nonprotein sulphydryl content. Different sensitivities in rat liver kidney. *Res. Comm. Chem. Pathol. Pharmacol.* **47**, 203–209 (1985).

71. T. D. Kinney, N. Kaufman, and J. V. Klavins. Deposition of iron in association with a periodic acid-Schiff (PAS)-positive material in the liver of ethionine-treated rats. *Lab. Invest.* **12**, 978–984 (1963).

72. J. E. Kinsella. Increased lipoperoxide content of orotic acid-induced fatty liver. *Biochim. Biophys. Acta* **137**, 205–207 (1967).

73. J. E. Kinsella. Protein and lipoperoxide levels in orotic acid-induced fatty liver. *Canad. J. Biochem.* **45**, 1206–1211 (1967).

74. N. Kocak-Tokez, M. Uytal, G. Aykac, A. Sivas, S. Yalcin, and H. Oz. Influence of acute ethanol administration on hepatic glutathione peroxidase and glutathione transferase activities in the rat. *Pharmacol. Res. Comm.* **17**, 233–239 (1985).

75. M. Koes, F. Ward, and S. Pennington. Lipid peroxidation in chronic ethanol-treated rats: in vitro uncoupling of peroxidation from reduced nicotine adenosine dinucleotide phosphate oxidation. *Lipids* **9**, 899–904 (1974).

76. D. R. Koop, E. T. Morgan, G. Tarr, and M. J. Coon. Purification and characterization of a unique isozyme of cytochrome-p-450 from liver microsomes of ethanol-treated rabbits. *J. Biol. Chem.* **257**, 8472–8480 (1982).

77. G. Krikun and A. I. Cederbaum. Effect of chronic ethanol consumption on microsomal lipid peroxidation. *FEBS Lett.* **208**, 292–296 (1986).

78. E. Laconi, P. M. Rao, S. Rajalakshmi, and D. S. R. Sarma. Posible mechanisms of liver tumor promotion in the rat, as revealed by orotic acid model, in *Models and Mechanisms in Chemical Carcinogenesis*, IV Sardinian Intern. Meeting, Alghero, October 23–27, 1987, p. 57A.

79. P. Lambelet, F. Saucy, and J. Löhger. Chemical evidence for interactions between vitamins E and C. *Experientia* **41**, 1384–1388 (1985).

80. T. B. Leonard, J. G. Dent, M. E. Graichen, O. Lyght, and J. A. Popp. Comparison of hepatic carcinogen initiation promotion systems. *Carcinogenesis* **3**, 851–856 (1982).

81. C. S. Lieber, D. P. Jones, J. M. Mendelson, L. M. De Carli. Fatty liver, hyperlipemia, and hyperuricemia produced by prolonged alcohol consumption despite adequate dietary intake. *Trans. Ass. Amer. Physicians* **76**, 289–301 (1963).

82. C. S. Lieber. Metabolic effects of ethanol on the liver and other digestive organs. *Clin. Gastroenterol.* **10**, 315–342 (1981).

83. C. S. Lieber. Ethanol metabolism and toxicity, in *Reviews in Biochemical Toxicity*, E. Hodgson, J. R. Bend, and R. M. Philpot, eds., Elsevier, New York, 1983, pp. 267–311.

84. E. R. Litov, O. H. Irving, J. F. Downey, and J. F. Tappel. Lipid peroxidation a mechanism involved in acute ethanol toxicity as demonstrated by "in vivo" pentane production in the rat. *Lipids* **13**, 305–307 (1978).

85. J. E. Loeb, C. Creuzet, O. Komano, and J. P. Boissel. A membrane-bound protein kinase from mouse liver stimulated by iron. *FEBS Lett.* **156**, 316–320 (1983).

86. B. Lombardi and S. Banni. Choline-devoid diet and hepatic cancerogenesis in the rat, in *Models and Mechanisms in Chemical Carcinogenesis*, IV Sardinian Intern. Meeting, Alghero, October 23–27, 1987, p. 119A.

87. C. M. MacDonald. The effects of ethanol on hepatic lipid peroxidation and on the activities of glutathione reductase and peroxidase. *FEBS Lett.* **35**, 227–230 (1973).

88. C. M. MacDonald, J. Dow, and M. R. Moore. A possible protective role for sulphydryl compounds in acute alcoholic liver injury. *Biochem. Pharmacol.* **26**, 1529–1531 (1977).

89. R. A. MacDonald. Experimental pigment cirrhosis: its production in rats by feeding a choline-deficient diet with excess iron. *Amer. J. Pathol.* **36**, 499–519 (1960).

90. R. A. MacDonald. 'Lifespan' of liver cells. Autoradiographic studies using tritiated thymidine in normal, cirrhotic, and partially hepatectomized rats. *Arch. Intern. Med.* **107**, 79–87 (1961).

91. R. Mazzanti, K. S. Srai, E. S. Debnam, A. M. Boss, and P. Gentilini. The effect of chronic ethanol consumption on iron absorption in rats. *Alcohol and Alcoholism* **22**, 47–53 (1987).

92. A. J. Montserrat, A. K. Ghoshal, W. S. Hartroft, and E. A. Porta. Lipoperoxidation in the pathogenesis of renal necrosis in choline-deficient rats. *Amer. J. Pathol.* **55**, 163–190 (1969).
93. A. Müller and H. Sies. Role of alcohol dehydrogenase activity and acetaldehyde in ethanol-induced ethane and pentane production by isolated perfused rat liver. *Biochem. J.* **206**, 153–156 (1982).
94. P. M. Newberne, J. L. V. de Camargo, and A. J. Clark. Choline-deficiency, partial hepatectomy, and liver tumors in rats and mice. *Toxicol. Pathol.* **2**, 95–106 (1982).
95. R. Nordmann, C. Ribière, and H. Rouach. Involvement of iron and iron-catalyzed free radical production in ethanol metabolism and toxicity. *Enzyme* **37**, 57–69 (1987).
96. A. B. Novikoff, P. S. Roheim, and N. Quintana. Changes in rat liver cells induced by orotic acid feeding. *Lab. Invest.* **15**, 27–33 (1966).
97. P. M. Novikoff and D. Edelstein. Reversal of orotic acid-induced fatty liver in rats by chlofibrate. *Lab. Invest.* **36**, 215–231 (1977).
98. P. M. Novikoff, P. S. Roheim, A. B. Novikoff, and B. S. Edelstein. Production and prevention of fatty liver in rats fed chlofibrate and orotic acid diets containing sucrose. *Lab. Invest.* **30**, 732–750 (1974).
99. R. Nunes, O. Beloqui, B. J. Potter, and P. D. Berk. Iron uptake from transferrin by isolated hepatocytes: effect of ethanol. *Biochem. Biophys. Res. Comm.* **125**, 824–830 (1984).
100. H. H. H. Oei, H. C. Zoganas, J. M. McCord, and S. W. Schaffer. Role of acetaldehyde and xanthine oxidase in ethanol-induced oxidative stress. *Res. Comm. Chem. Pathol. Pharmacol.* **51**, 195–203 (1986).
101. J. E. Packer, T. F. Slater, and R. L. Wilson. Direct observation of a free radical interaction between vitamin E and vitamin C. *Nature* **278**, 737, 738 (1979).
102. S. Parodi, C. Balbi, M. Taningher, M. Pala, F. Marchesini, R. Bordone, D. S. R. Sarma, and L. Santi. 1% orotic acid in the diet, as a promoter of rat liver carcinogenesis, is inducing DNA fragmentation or changes in chromatin conformation? in *Models and Mechanisms in Chemical Carcinogenesis*, IV Sardinian Intern. Meeting, Alghero, October 23–27, 1987, pp. 48,49A.
103. M. I. R. Perera, J. M. Betshart, M. A. Virji, and H. Shinozuka. Free radical injuries and liver tumor promotion, in *Models and Mechanisms in Chemical Carcinogenesis*, IV Sardinian Intern. Meeting, Alghero, October 23–27, 1987, p. 10A.
104. M. I. R. Perera, A. J. Demetris, S. L. Katyal, and H. Shinozuka. Lipid peroxidation of liver microsome membranes induced by choline-deficient diets and its relationship to the diet-induced promotion of the induction of γ-glutamyltranspeptidase positive foci. *Cancer Res.* **45**, 2533–2538 (1985).
105. T. J. Peters, M. J. O'Connell, S. Venkatesan, and R. J. Ward. Evidence for free radical-mediated damage in experimental and human alcoholic liver disease, in *Free Radicals, Cell Damage, and Disease*, (C. Rice-Evans, ed.), Richelieu Press, London, 1986, pp. 99–110.
106. G. Poli, E. Albano, F. Biasi, G. Cecchini, R. Carini, G. Bellomo, and M. U. Dianzani. Lipid peroxidation stimulated by carbon tetrachloride or iron and hepatocyte death: protective effect of vitamin E, in *Free Radicals in Liver Injury*, (G. Poli, K. H. Cheeseman, M. U. Dianzani, and T. F. Slater, eds.), IRL Press, Oxford, 1986, pp. 207–215.
107. G. Poli, E. Albano, E. and M. U. Dianzani. The role of lipid peroxidation in liver damage. *Chem. Phys. Lipids* **44**, 117–142 (1987a).

108. G. Poli, E. Albano, A. Tomasi, K. H. Cheeseman, E. Chiarpotto, M. Parola, M. E. Biocca, T. F. Slater, and M. U. Dianzani. Electron spin resonance studies in isolated hepatocytes treated with ferrous or ferric iron. *Free Rad. Res. Comm.* **3**, 251–255 (1987b).

109. G. Poli, E. Chiarpotto, E. Albano, F. Biasi, G. Cecchini, and M. U. Dianzani. Iron overload: experimental approach using rat hepatocytes in single cell suspension, in *Frontiers in Gastrointestinal Research*, (M. U. Dianzani and P. Gentilini eds.), Karger, Basel, 1987c, pp. 38–49.

110. G. Poli, M. U. Dianzani, K. H. Cheeseman, T. F. Slater, J. Lang, and H. Esterbauer. Separation and characterization of the aldehydic products of lipid peroxidation stimulated by carbon tetrachloride or ADP-iron in isolated rat hepatocytes and rat liver microsomal suspensions. *Bochem. J.* **227**, 629–638 (1985).

111. L. A. Pottenger and G. S. Getz, Serum lipoproteins accumulation in the livers of orotic acid-fed rats. *J. Lipid Res.* **12**, 450–456 (1971).

112. K. N. Rao. Communications in *Models and Mechanisms in Chemical Carcinogenesis*, IV Sardinian Intern. Meeting, Alghero, October 23–27, 1987.

113. P. M. Rao, K. Nagamine, R. K. Ho, C. Laurier, S. Rajalakshmi, and D. S. R. Sarma. Dietary orotic acid enhance the incidence of y-glutamyl-transpeptidase positive foci in rat liver induced by chemical carcinogens. *Carcinogenesis* **4**, 1541–1545 (1983).

114. R. O. Recknagel and A. K. Ghoshal. Lipoperoxidation as vector in carbon tetrachloride hepatotoxicity. *Lab. Invest.* **15**, 132–156 (1966).

115. R. O. Recknagel and A. K. Ghoshal. New data on the question of lipoperoxidation in carbon tetrachloride poisoning. *Exp. Mol. Pathol.* **5**, 108–117 (1966).

116. D. C. Reitz. A possible mechanism for the peroxidation of lipids due to chronic ethanol ingestion. *Biochim. Biophys. Acta* **380**, 145–154 (1975).

117. H. Remmer, D. Albrecht, and H. Kappus. Lipid peroxidation in isolated hepatocytes from rats ingesting ethanol chronically. *Naunyn-Schmiedeberks Arch. Pharmakol.* **298**, 107–113 (1977).

118. G. W. Richter. Effects of cyclic starvation-feeding and of splenectomy on the development of hemosiderosis in rat liver. *Am. J. Pathol.* **74**, 481–506 (1974).

119. G. W. Richter. The iron-loaded cell. The cytopathology of iron storage. A review. *Am. J. Pathol.* **91**, 363–404 (1978).

120. M. A. Rossi, A. Giordano, and M. U. Dianzani. Effetto in vitro dell'etanolo dell'acetaldeide sulla perossidazione lipidica e sul contenuto di glutatione in omogenato di fegato. *S.I.B.S.* **LXII**, 1–6 (1986).

121. W. T. Roubal and A. L. Tappel. Damage to proteins, enzymes, and aminoacids by peroxidizing lipids. *Arch. Biochem. Biophys.* **113**, 5–8 (1966).

122. W. T. Roubal and A. L. Tappel. Polymerization of proteins induced by free-radical lipid peroxidation. *Arch. Biochem. Biophys.* **113**, 150–155 (1966).

123. S. M. Sabesin, S. Frase, and J. B. Ragland. Accumulation of nascent lipoproteins in rat hepatic Golgi during induction of fatty liver by orotic acid. *Lab. Invest.* **37**, 127–135 (1977).

124. M. Saito, L. A. Morehouse, and S. D. Aust. Transferring-dependent lipid peroxidation. *J. Free Rad. in Biol. Med.* **2**, 99–105 (1986).

125. M. A. Sells, S. L. Katyal, S. Sell, H. Shinozuka, and B. Lombardi. Induction of foci of fed a choline-deficient diet. *Br. J. Cancer* **40**, 274–283 (1979).

126. S. Shaw, E. Jayatilleke, and C. S. Lieber. The effect of chronic alcohol feeding on lipid peroxidation in microsomes; lack of relationship to hydroxyl radical generation. *Biochem. Biophys. Res. Comm.* **118**, 233–238 (1984).

127. S. Shaw, E. Jayatilleke, W. A. Ross, E. R. Gordon, and C. S. Lieber. Ethanol-induced lipid peroxidation: potentiation by long-term alcohol feeding and attenuation by methionine. *J. Lab. Clin. Med.* **98**, 415–424 (1981).

128. S. Shaw, K. D. Rubin, and C. S. Lieber. Depressed hepatic glutathione and increased diene conjugates in alcohol liver disease. *Dig. Dis. Science* **28**, 585–589 (1983).

129. Y. Shigeta, F. Nomura, S. Iida, M. A. Leo, M. R. Felder, and C. S. Lieber. Ethanol metabolism in vivo in the microsomal ethanol-oxidizing system in deermice lacking alcohol dehydrogenase (ADH). *Biochem. Pharmacol.* **33**, 807–814 (1984).

130. H. Shinozuka, C. Gupta, A. Hattori, J. M. Betschart, and M. A. Virji. Choline-deficiency, lipid peroxidation, liver cell surface receptor alteration, and liver tumor promotion, in *Models and Mechanisms in Chemical Carcinogenesis*, IV Sardinian Intern. Meeting, Alghero, Ocotober 23–27, 1987. p. 124A.

131. H. Shinozuka and S. L. Katyal. Pathology of choline deficiency, in *Nutritional Pathology* (H. Sidransky ed.), Marcel Decker, New York, 1985, pp. 279–320.

132. H. Shinozuka and B. Lombardi. Synergistic effect of a choline-devoid diet and phenobarbital in promoting the emergence of foci of γ-glutamyl-transpeptidase-positive hepatocytes in the liver of carcinogen-treated rats. *Cancer Res.* **40**, 3846–3849 (1980).

133. H. Sies, O. R. Koch, E. Martino, and A. Boveris. Increased biliary glutathione disulphide release in chronically ethanol-treated rats. *FEBS Lett.* **103**, 287–290 (1979).

134. T. F. Slater. Necrogenic action of carbon tetrachloride in the rat: a speculative mechanism based on activation. *Nature* **209**, 36–40 (1966).

135. T. F. Slater. Free radical mechanisms, in *Tissue Injury*, Pion Press, London, 1972, pp. 1–283.

136. T. F. Slater. Free radical mechanisms in tissue injury. *Biochem. J.* **222**, 1–15 (1984).

137. A. Slivka, J. Kang, and G. Cogen. Hydroxyl radicals and the toxicity of oral iron. *Biochem. Pharmacol.* **35**, 553–556 (1986).

138. H. Speisky, D. Bunout, H. Orrego, H. G. Giles, A. Gunasekara, and Y. Israel. Lack of changes in diene conjugates levels following ethanol-induced glutathione depletion or hepatic necrosis. *Res. Comm. Chem. Pathol. Pharmacol.* **48**, 77–90 (1985).

139. S. B. Standerfer and P. Handler. Fatty liver induced by orotic acid feeding. *Proc. Soc. Exp. Biol. Med.* **90**, 270–271 (1955).

140. P. E. Starke, J. D. Gilbertson, and J. Farber. Lysosomal origin of the ferric iron required for cell killing by hydrogen peroxide. *Biochem. Biophys. Res. Comm.* **133**, 371–379 (1985).

141. E. Stege. Acetaldehyde-induced lipid peroxidation in isolated hepatocytes. *Res. Comm. Chem. Pathol. Pharmacol.* **36**, 287–297 (1982).

142. A. Takada, F. Ikegami, Y. Okumura, Y. Hasumura, R. Kanayama, and J. Takeuchi. Effect of alcohol on the liver of rats. III. The role of lipid peroxidation and sulphydryl compounds in ethanol-induced liver injury. *Lab. Invest.* **23**, 421–428 (1970).

143. T. Takagi, J. Alderman, J. Gellet, and C. S. Lieber. Assessment of the role of non-ADH ethanol oxidation in vivo and in hepatocytes free deermice. *Biochem. Pharmacol.* **35**, 3601–3606 (1986).

144. M. D. Takeuchi, A. Takada, K. Ebata, G. Sawae, and Y. Okumura. Effect of alcohol on the liver of rats. I. Effect of a single intoxicating dose of alcohol

on the livers of rats fed a choline-deficient diet or a commercial ration. *Lab. Invest.* **19,** 211–217 (1968).

145. K. H. Tan, D. J. Meyer, and B. Ketterer. Lipid peroxidation in choline-methionine deficiency. *Free Rad. Res. Comm.* **3,** 273–278 (1987).

146. A. Tomasi, E. Albano, M. U. Dianzani, and V. Vannini. Metabolic activation of 1,2-dibromoethane to a free radical intermediate by rat liver microsomes and isolated hepatocytes. *FEBS Lett.* **160,** 191–194 (1983).

147. M. V. Torrielli, M. U. Dianzani, and G. Ugazio. Behavior of lipoperoxidation in rat liver during orotic acid treatment. *Life Sci.* **10,** 99–111 (1971).

148. M. V. Torrielli, L. Gabriel, and M. U. Dianzani. Ethanol-induced hepatotoxicity: experimental observations on the role of lipid peroxidation. *J. Pathol.* **126,** 11–25 (1978).

149. M. V. Torrielli and G. Ugazio. Effect of DPPD on the orotic acid-induced fatty liver in the rat. *Life Sci.* **9,** 1–7 (1970).

150. G. Ugazio, L. Gabriel, and E. Burdino. Osservazioni sperimentali sui lipidi accumulati nel fegato di ratti alimentati con dieta colino-priva. *Lo Sperimentale* **117,** 1–17 (1967).

151. F. J. G. M. Van Kuijk, D. W. Thomas, R. J. Stephens, and E. A. Dratz. Occurrence of 4-hydroxy-alkenals in rat tissues determined as pentafluorobenzyloxime derivatives by gas chromatography. *Biochem. Biophys. Res. Comm.* **139,** 144–149 (1986).

152. A. Valenzuela, V. Fernandez, and L. A. Videla. Hepatic and biliary levels of glutathione and lipid peroxides following iron overload in the rat: effect of simultaneous ethanol administration. *Toxicol. Appl. Pharmacol.* **70,** 87–95 (1983).

153. S. Vasudevan, E. Laconi, P. M. Rao, S. Rajalakshmi, and D. S. R. Sarma. Can metabolic disturbances generate tumor promoters? In *Models and Mechanisms in Chemical Carcinogenesis,* IV Sardinian Intern. Meeting, Alghero, October 23–27, 1987. p. 67A.

154. L. A. Videla, V. Fernandez, A. De Marinis, N. Fernandez, and A. Valenzuela. Lipoperoxidative pressure and glutathione status following acetaldehyde and aliphatic alcohols pretreatments in the rat. *Biochem. Biophys. Res. Comm.* **104,** 965–970 (1982).

155. L. A. Videla, V. Fernandez, G. Ugarte, and A. Valenzuela. Effect of acute ethanol intoxication on the content of reduced glutathione of the liver in relation to its lipoperoxidative capacity in the rat. *FEBS Lett.* **III,** 6–10 (1980).

156. L. A. Videla, C. G. Fraga, O. R. Koch, and A. Boveris. Chemiluminescence of the *in situ* rat liver after acute ethanol intoxication. Effect of (+)-cyanidanol-. *Biochem. Pharmacol.* **32,** 2822–2825 (1983).

157. L. A. Videla and A. Valenzuela. Alcohol ingestion, liver glutathione, and lipoperoxidation: metabolic interrelations and pathological implications. *Life Sci.* **31,** 2395–2407 (1982).

158. H. G. Windmueller. Depression of plasma lipids in the rat by orotic acid and its reversal by adenine. *Biochem. Biophys. Res. Comm.* **II,** 496–500 (1963).

159. H. G. Windmueller. An orotic acid-induced, adenine-reversed inhibition of hepatic lipoprotein secretion in the rat. *J. Biol. Chem.* **259,** 530–537 (1964).

160. H. G. Windmueller and R. J. Levy. Total inhibition of hepatic lipoprotein production in the rat by orotic acid. *J. Biol. Chem.* **242,** 2246–2254 (1967).

161. K. Yagi. Lipid peroxides, in *Biology and Medicine,* Academic, New York, 1982, pp. 1–351.

162. S. Yokoyama, M. A. Sells, T. V. Reddy, and B. Lombardi. Hepatocarcinogenic and promoting action of a choline-devoid diet in the rat. *Cancer Res.* **45,** 2834–2842 (1985).
163. M. Yoshioka and T. Kanada. Studies on the toxicity of the autoxidized oils. III. The toxicity of hydroperoxyalkenals. *Yukagaku* **23,** 321–326 (1974).
164. S. P. Young, S. Roberts, A. Bomford. Intracellular processing of transferrin and iron by isolated rat hepatocytes. *Biochem. J.* **232,** 819–823 (1985).

From: *Trace Elements, Micronutrients, and Free Radicals* • Ed.: I. E. Dreosti • ©1991 The Humana Press Inc.

CHAPTER 5

Essential Trace Elements in Antioxidant Processes

SHERI ZIDENBERG-CHERR AND CARL L. KEEN

ABSTRACT

The antioxidant defense system is comprised of a number of interconnecting, and overlapping, components that include both enzymatic and nonenzymatic factors. The trace elements Cu, Zn, and Mn are critical components for a number of these processes, and a deficiency of any one of these elements can result in an impairment of the functioning of the overall antioxidant system. This impairment can be physiologically significant, particularly if the animal is exposed to environmental challenges that increase the production of oxygen radicals over normal physiological levels.

1. INTRODUCTION

The formation of highly reactive oxygen-containing molecular species is a normal consequence of cellular metabolism. For example, sources of superoxide anion ($O_2{}^{\cdot-}$) include: the actions of cytochrome oxidase, xanthine oxidase, the autooxidation of numerous molecules, including catecholamines, leucoflavins, and tetrahydropterins, and the autooxidation of a number of proteins, including hemoproteins. Additionally, such species are generated as a result of the metabolism of a number of compounds, including xenobiotics, organic solvents, alcohol, and ozone. Numerous cellular processes are dependent on oxygen radical species, and they are critical for neutrophil, monocyte, and macrophage function. However, an excess of these species can result in peroxidative damage to cell membranes, proteins, and nucleic acids. Thus, in order to maintain tissue integrity, there must be a critical balance between the production of reactive free radicals, their use in essential pathways, and the clearance of excess radicals by the "antioxidant system."

The "antioxidant system" includes a number of enzymes and low mol wt compounds, many of which are, or are dependent on, essential nutrients. These include vitamin E (tocopherol), vitamin C (ascorbic acid), beta-carotene, zinc (Zn), copper (Cu), manganese (Mn), iron (Fe), and selenium (Se). The vitamins are not thought to be dependent on other factors to participate in free radical defense, whereas the metals exert their action as antioxidants primarily via their incorporation into specific enzymes. The metals can also have either prooxidant or antioxidant effects when complexed with low molecular weight ligands, or are present as the ionic species.

The dietary intake of the nutrients mentioned above can modulate the activity of the antioxidant defense system, and thus impact on the degree of protection provided to the cell or tissue against oxidative reactions. Thus, a dietary deficiency of one of these defense components can impair the ability of the organism to defend itself against excessive free radical damage and its associated pathologies. In this paper, we will focus on the role of the trace elements Cu, Zn, and Mn as critical components of the body's antioxidant defense system. The biochemical lesions that occur as a consequence of a deficiency of one or more of the above elements functioning in the antioxidant defense system will be emphasized. Owing to space constraints, Se will not be discussed. For a comprehensive review of this element, *see* reference (48).

2. COPPER

Copper is an essential nutrient found in all mammalian tissues, where it is primarily bound to proteins or other organic molecules. Copper exhibits mono- and divalence, and forms water-soluble cationic simple salts. Copper metalloenzymes are involved in oxidation–reduction reactions with O_2 acting as the electron acceptor, and Cu is involved in the electron transfer. In tissues, Cu^+ ions are oxidized to Cu^{2+} by peroxide, which, when bound to -SH-containing compounds, can reversibly react to form Cu^+ and disulfide. Typical soft tissue Cu concentrations range from 50–150 μM and serum levels range from 10–30 μM.

Although dietary Cu deficiency is thought to be rare in humans, it is a fairly common occurrence in sheep and cattle, which consume forage low in this element (36). Typical deficiency signs include anemia, impaired immunocompetence, altered neurotransmitter function, skeletal defects, vascular defects, emphysemic type lung defects, and death caused by cardiac rupture or aortic aneurysms (36). Below, we will focus on the contribution of Cu to the antioxidant system and the events that occur as a result of its deficiency.

2.1. Cu and the Antioxidant System

Four antioxidant enzymes whose activities have been reported to be reduced under conditions of Cu deficiency are (1) Cu,Zn superoxide dismutase (Cu,ZnSOD), an enzyme that catalyzes the conversion of O_2^{-} to H_2O_2; (2) ceruloplasmin (Cp; ferroxidase I), a Cu-containing enzyme that can act as a circulating free radical scavenger by virtue of its SOD activity, and also may prevent redox cycling as a result of its ability to maintain iron in the ferric form; (3) catalase; and (4) selenium-dependent glutathione peroxidase (GSH-Px), that converts hydroperoxy compounds to hydroxyl compounds. Where Cu deficiency-induced reductions in tissue Cu,ZnSOD and plasma Cp activity are consistent findings in most laboratories (Table 1), reports on the effect of Cu deficiency on GSH-Px activity are variable. It should be noted that discrepancies in the literature regarding the effect of Cu deficiency on the antioxidant defense systems may be related in part to the varying degrees of the Cu deficiency induced. For example, when rats were fed either evaporated milk- or powdered milk-based Cu-deficient diets for 8–10 wk, liver Cu concentrations were 8% and 35% of their controls, respectively. Whereas both situations resulted in similar reductions in Cu,ZnSOD and Cp activity, only the rats fed the evaporated milk-based diet showed lower than normal GSH-Px activity (37). Thus, the effect of Cu deficiency on GSH-Px may represent a late, rather than early, biochemical lesion. It is also important to note that when the activity, or concentration, of a component of the free radical system is affected, it is likely to have an influence on other components of the system.

Whereas some of the explanations underlying the effect of Cu deficiency on the antioxidant system are evident, others are less clear. Since Cu,Zn-SOD and Cp are Cu-containing proteins, it is predictable that the activity of these enzymes will be influenced by tissue Cu concentrations. It should be noted that whereas the tissue activities of these enzymes are reduced under deficiency conditions, the synthesis of the apoenzymes may not be inhibited (11).

In contrast to Cu,ZnSOD and Cp, the mechanisms by which Cu influences the activity of GSH-Px and catalase are less clear. With regard to the effect of Cu deficiency on catalase activity, anemia and high liver Fe concentrations in Cu-deficient rats have been reported in a number of studies. It has been suggested that these phenomenon are related to defects in Fe transport and, possibly, in heme iron incorporation (16,17). Since catalase is a heme-containing enzyme, it is possible that the Cu deficiency is directly affecting its synthesis, although other secondary mechanisms cannot be ruled out.

Table 1
Effects of Cu Deficiency on the Antioxidant Defense System in Liver

Species	Sex	Diet (Cu ppm)	Duration	CuZnSOD	GSH-Px	Catalase	GSH	LPO[1]	Ref
Rat	M	0.02[4]	8 wk	→	→	–[3]	–	ND[2]	37
Rat	M	0.02[4]	8 wk	→	→	–	–	–	33
Rat	M	0.2	7 wk	→	→	–	←	–	1
Mouse	M	0.2	6 wk	→	–	–	←	–	56
Rat	M	<1	4 wk	→	→	→	–	←	3
Rat	M	1.2	11 wk	→	→	–	–	←	23
Rat	M	0.8	8 wk	→	→	–	–	→	4
Rat	M	0.8	6 wk	→	–	–	ND	←	39
Rat	F	0.8	Gest, 6 wk	→	ND	→	→	–	45
Mouse	F/M	0.5	Gest, 12 d	–	↓/ND[5]	–	–	–	40

[1]Includes measurements of TBARS, lipid hydroperoxides, and hydrocarbon gases as indices of lipid peroxidation.
[2]No differences.
[3]Not measured.
[4]0.02 mg/mL of liquid diet.
[5]Response was strain dependent.

There have been a number of studies that have suggested that interactions between Cu and Se are responsible, at least in part, for the effect of Cu deficiency on GSH-Px activity. For example, when weanling male rats were fed evaporated milk-based Cu-deficient diets, they had low liver Cu,ZnSOD activity, and low liver and lung GSH-Px activity; the activity of non-Se-dependent GSH-Px was unaffected in these tissues. The Cu-deficient rats exhibited an increase in the fecal excretion of Se that was independent of biliary flow. When the Cu-deficient diet was supplemented with Se (0.1 μg Se/mL diet), the activity of GSH-Px was increased from 30 to 70% of control levels (33). These results support the idea that changes in Se metabolism play a role in the Cu deficiency-induced reduction in GSH-Px activity.

Given the observation that Cu deficiency can affect the activities of GSH-Px, it is not surprising that GSH and GSSG levels can be altered by a deficiency of this element. Allen et al. (1) reported that when rats were fed Cu-deficient diets (0.2 ppm) for 7 wk, hepatic GSH concentrations were increased by approx 50%, although GSSG concentrations were unaffected. Similar to the above, we observed that when weanling mice were fed Cu-deficient diets (0.2 ppm) for 6 wk, hepatic GSH concentrations were 150% higher than in controls (56). Since GSH-Px requires GSH as substrate in the conversion of lipid peroxides to lipid alcohol, the increased levels of GSH in the Cu-deficient animal may reflect the low GSH-Px activity. Alternatively, since GSH also functions independent of GSH-Px in the protection from free radical damage, Cu-deficient animals may have an increased need for GSH, owing to increased substrate utilization, which could result in an increased synthesis of GSH. There is precedence for this idea, in that increased hepatic GSH synthesis has been reported in rats exhibiting low hepatic GSH-Px activity owing to Se deficiency (29). Similarly, rats chronically treated with ethanol have elevated hepatic levels of GSH concomitant with increased levels of lipid peroxides (27).

2.2. Functional Significance of Cu Deficiency-Induced Changes in the Free Radical Defense System

Whereas it is clear that dietary Cu deficiency can result in a reduction in the activity of Cu,ZnSOD, Cp, and GSH-Px, there is considerable debate over the significance of these reductions. However, evidence is accumulating that Cu deficiency and the resulting impairment in enzymatic and nonenzymatic defense mechanisms can increase the oxidative stress imposed upon the animal. Balevska et al. (3) reported that when adult rats were fed Cu-deficient diets for 4 wk, there was an increase in hepatic mitochondrial and microsomal lipid peroxidation (as estimated by the formation of lipid hydroperoxides), which was attributed to Cu deficiency-induced reductions in GSH-Px. Subsequent to the work by Balevska et al., other investigators have reported both increased and decreased levels of lipid peroxidation in Cu-deficient animals

(*4,23,39*). For example, Paynter (*39*) did not find evidence of increased lipid peroxidation (as estimated by thiobarbituric acid reacting substances (TBARS)) in rats that were fed diets deficient in Cu but adequate in Se and Mn. However, the production of hepatic TBARS was increased in rats fed diets deficient in both Cu and Mn, or deficient in both Cu and Se to an extent greater than that in rats fed either Mn-deficient:Cu-adequate or Se-deficient:Cu-adequate diets. This paper underscores the potential interaction of these nutrients in the antioxidant defense system.

The increased level of lipid peroxidation associated with Cu deficiency has been suggested to be physiologically significant. Inadequate Cu,ZnSOD activity with subsequent peroxidation of erythrocyte membrane lipids has been proposed as one explanation for the increased osmotic fragility and shortened life-span of erythrocytes from Cu-deficient animals. Erythrocytes from Cu-deficient rats also display a number of changes in lipid composition, which may play a role in the altered lifespan of these red blood cells, including an increase in total cholesterol and phospholipid concentrations, and an increase in phospholipid–malondialdehyde adducts, indicative of lipid peroxidation. Jain and Williams (*31*) have proposed that both increased lipid peroxidation, because of reduced Cu,ZnSOD activity, as well as altered lipid synthesis caused by reduced plasma lecithin-cholesterol acyl transferase activity, contribute to the shortened life-span of the Cu-deficient erythrocyte and the resultant anemia that is commonly noted in Cu-deficient animals.

In addition to erythrocyte membranes, alterations in membrane lipid and fatty acid composition have been reported in cellular and subcellular membranes from livers of Cu-deficient rats. Specific changes with respect to microsomal membrane fatty acid composition can include reductions in the concentration of polyunsaturated fatty acids (18:2, 20:4, 22:6) and increases in the concentration of monounsaturated fatty acids (16:1 and 18:1); saturated fatty acid (16:0 and 18:0) concentrations typically remain constant (*4*). It is significant to note that these changes should result in a reduction in the peroxidizability of the membranes and consistent with this, despite marked reductions in liver Cu,ZnSOD activity, it has been reported that microsomes from Cu-deficient rats show an increased resistance to lipid peroxidation relative to controls (*4*). It is interesting to note that, similar to Cu-deficient cells, Morris hepatoma cells are characterized by a low CuZnSOD activity, and a high ratio of saturated/unsaturated fatty acids in the membrane. Baroli et al. (*4*) has suggested that inadequate SOD activity similar to that seen with Cu deficiency may be the cause of the modifications of microsomal membrane fatty acid composition observed in tumor cells. According to this idea, Cu deficiency-impaired antioxidant systems generate high oxygen radical concentrations that trigger the synthesis of a more saturated pattern of fatty acids comprising phospholipids that are relatively insensitive to further peroxidation. Variations in the degree of fatty acid changes in the response to Cu deficiency may contribute to the variable results on Cu deficiency-induced effects on lipid peroxidation in the literature.

At present there is little information concerning the functional significance of low plasma Cp activity in Cu-deficient animals with regard to free radical defense mechanisms. However, given its ability to scavenge superoxide anion radicals (26), the effect of low levels of this enzyme on circulating catecholamine metabolism and vessel wall peroxidation merits investigation.

To further evaluate the functional significance of the compromised antioxidant defense system observed in the Cu-deficient animal, the influence of Cu deficiency on an animals' response to agents that generate free radicals can be studied. For example, we have used the drug adriamycin (ADR; doxorubicin) as a probe to assess the response of Cu-deficient mice to increased free radical stress. ADR is a chemotherapeutic drug whose value is compromised by the cardiotoxicity associated with its long-term use. One mechanism that has been proposed to contribute to this side effect is that peroxidative reactions are induced by ADR treatment, since its metabolism results in the generation of free radicals (49). When Cu-sufficient and Cu-deficient mice were treated with ADR, hepatic GSH concentrations were increased, the highest concentrations occurring in the Cu-deficient mice. A different response was observed in cardiac tissue; ADR treatment resulted in an increase in GSH concentration in Cu-sufficient mice, whereas Cu-deficient mice exhibited lower GSH concentrations relative to controls (56). The cause for low levels of cardiac GSH in Cu-deficient mice relative to controls is unclear; since ADR will induce a reduction in food intake, it may result in an inadequate supply of substrates for GSH synthesis, and utilization by the heart. Liver is the main site for synthesis of GSH, and whereas Cu deficiency may result in increased hepatic synthesis of GSH, subsequent export to the plasma for utilization by other tissues may be hindered in Cu-deficient animals.

In addition to the use of ADR as a probe to induce oxidative stress, we have investigated the effects of low tissue Cu concentrations on an animal's response to ethanol administration. When pregnant rats were fed ethanol-containing diets that were adequate in all nutrients throughout gestation, fetal liver Cu,ZnSOD activity was increased at day 21 of gestation (55). Similar findings have been reported by Dreosti and Record (19), with the exception that these authors noted an increase in the activity of fetal liver MnSOD, rather than Cu,ZnSOD. The above suggests that exposure to ethanol *in utero* may increase the oxygen radical concentration in fetal liver, causing the fetus to require increased antioxidant protection. In contrast to fetuses with adequate Cu stores, fetuses containing low Cu concentrations (60% of controls) were not able to respond to metabolites of ethanol by increasing liver Cu,ZnSOD activity.

activity. This suggests that fetuses whose antioxidant protection may be compromised via diet, metabolic disturbance, or drug treatment, may be more sensitive to the effects of ethanol *in utero*.

In both chronic and acute ethanol-treated nonpregnant rats, the activity of hepatic SOD and GSH-Px have been reported to be altered. However, in this situation, it is MnSOD rather than Cu,ZnSOD that increases (20). Similar to the ethanol-exposed fetus, this phenomenon is probably a protective response to the generation of free radicals resulting from ethanol metabolism. Similar to ADR and alcohol, other potential free radical generators whose toxicity has been reported to be increased under conditions of Cu deficiency include carbon tetrachloride and hyperoxia (37,42). Thus, whereas this is still an evolving area of research, current evidence supports the idea that the toxicity of compounds that produce free radicals via their metabolism are enhanced under conditions of Cu deficiency owing to an impairment in the antioxidant system.

3. ZINC

Zinc forms stable water soluble salts exhibiting a valance of $+2$, except at elevated pH, where it forms insoluble $Zn(OH)_2$. Zinc readily complexes with amino acids, nucleotides, peptides, and proteins, and has a high affinity for thiol and hydroxyl groups, and for ligands containing N as a donor. In contrast to Cu and Mn, Zn does not undergo oxidation–reduction reactions, and thus, it is not involved in electron transfer reactions.

In soft tissues, Zn concentrations typically range from 300 to 600 μM, and in plasma from 10 to 25 μM. Severe zinc deficiency can result in anorexia, growth retardation, abnormal immune function, impaired reproductive capacity, skin lesions, and behavioral defects (14). Whereas severe Zn deficiency may be a rare occurrence in human populations, marginal Zn deficiency can be a relatively common problem in both humans and domestic animals, depending on their diet and physiological condition (36).

That a deficiency of Zn results in numerous biochemical and structural defects is not surprising, considering that Zn ions are components of over 200 enzymes. In addition to the requirement of Zn for enzymatic function, this metal has a number of nonenzymatic roles essential for normal tissue growth and maintenance. As a result of its chelation properties, Zn serves a role in the structure and function of biomembranes; indeed, it has been suggested that many of the pathological signs of Zn deficiency may be explained by a general membrane defect which, as described below, may be related in part to excessive lipid peroxidation (7).

3.1. Zinc and the Antioxidant Defense System

There are at least three mechanisms by which Zn is directly involved in antioxidant defense. The first involves its presence in the enzyme Cu,ZnSOD. Whereas the enzyme has an absolute requirement for Zn, under most conditions, there is normal activity of this enzyme, even under conditions of severe Zn deficiency (45) (Table 2). Thus, this component of the antioxidant defense system does not appear to be susceptible to Zn deficiency. Zinc also functions as an antioxidant through its interactions with specific cell surface components. Through this type of interaction, the molecular conformation of the membrane may be altered and/or the metal may compete with redox active metals for membrane binding sites, thereby preventing formation of hydroxyl radicals resulting from redox cycling (13). A third mechanism by which Zn may function in free radical defense is through its association with the sulfur-rich, low mol wt protein, metallothionein (mt). Zn-mt can be induced by a number of conditions that increase oxidative stress, including hyperoxia, ionizing radiation, and exposure to xenobiotics (41). It has been suggested that the Zn thiolate clusters present in this protein are efficient hydroxyl radical scavengers (46). An additional thiol compound influenced by Zn status is GSH; its synthesis is increased in liver from Zn-deficient rats (30). In contrast to the lack of an effect of Zn deficiency on Cu,ZnSOD activity, the contribution that Zn ions make in terms of membrane stabilization, as well as the putative role of Zn-mt as a cellular antioxidant, is dependent on the Zn status of the individual.

3.2. Functional Significance of Zn Deficiency-Induced Changes in the Free Radical Defense System

Some of the initial research focusing on the association between Zn deficiency and lipid peroxidation was provided by Bettger et al. (6), who demonstrated that low plasma Zn was associated with increased erythrocyte fragility. When erythrocytes from control animals were incubated in media containing different cations, Cu (16-80 μM) increased fragility; this hemolytic effect of Cu was associated with an increase in lipid peroxidation as assessed by TBARS. When Zn (16 μM) was added along with Cu, the hemolytic effect of Cu was reduced; however, Zn did not counteract the increase in TBARS production. These results suggest that although Zn may not prevent Cu-induced peroxidation, it has the ability to stabilize membranes once the event has occurred.

Since Zn deficiency is characterized by low plasma Zn and often associated with increased plasma Cu concentrations, the interaction between Zn and Cu mentioned above are of physiological significance. Although tissue Zn concentrations are not markedly affected in Zn deficiency, specific cellular fractions do exhibit reduced Zn concentrations.

Table 2
Effects of Zn Deficiency on the Antioxidant Defense System in Liver

Species	Sex	Diet	Duration	CuZnSOD	GSH-Px	Catalase	GSH	LPO[1]	Ref
		(Zn ppm)							
Rat	M	<1	6 wk	ND[3]	ND	↓	↑	–[2]	45
Rat	M	2.0	1 wk	–	–	–	↑	–	30
Rat	M	1.2	7	–	–	–	–	↑	44
Mouse	F	0.5	3 wk	–	–	–	–	↑	9
Rat	M/F	0.5	Gest	–	–	–	–	↑	21

[1] Includes measurements of TBARS, lipid hydroperoxides, and hydrocarbon gases as indices of lipid peroxidation.
[2] Not measured.
[3] No differences.

For example, erythrocytes from Zn-deficient rats contain less membrane-bound Zn than those from controls, whereas total hemolysate Zn concentrations remain unaffected (7). Thus, the increased fragility of erythrocytes in Zn-deficient animals may be a result of Cu-induced lipid peroxidation as well as a reduction in the membrane stabilization typically afforded by Zn. However, a critical determinant of the actual ability of Cu to induce free radical events is the form in which it is found. Typically, in conditions of Zn deficiency, the elevation in plasma Cu is related primarily to Cp-bound Cu. Information on the ability of Cu in this form to contribute to lipid peroxidation, as well as information on the other molecular species of Cu that are increased in Zn deficiency, are needed to better understand the physiological significance of this Zn–Cu interaction.

A common model system in which to study the role of Zn on membrane stabilization, and how its deficiency influences membrane lipid peroxidation and integrity, is through the use of isolated erythrocyte membranes. Using this system, Jay et al. (32) measured membrane fluidity in erythrocyte ghosts from rats fed Zn-deficient diets for 5 wk following weaning. Assessment of fluidity using the spin label 5-NS demonstrated that as Zn deficiency progresses, membrane fluidity increases. To determine whether the effect of Zn was occurring at the membrane surface, a spin label specific for sialic acid residues was utilized. Based on results from these experiments, it was concluded that Zn deficiency induced a conformational change in cell-surface glycoconjugates.

In addition to erythrocyte membrane lipid peroxidation, Zn deficiency can result in increased lipid peroxidation in other tissues and subcellular organelles, including liver mitochondria and microsomes (9,44), lung microsomes (8), and maternal and fetal liver (21).

Similar to the phenomenon that occurs in Cu deficiency, increased tissue Fe concentrations are a common feature of Zn deficiency (36). This lends support to the concept that one antioxidant function of Zn is to prevent the accumulation of membrane Fe and its subsequent free radical-promoting effects. Whereas the increase in microsomal peroxidation may be caused by the lack of Zn as a stabilizing factor and/or an "inhibitor" of low mol wt Fe localization in the membrane, it should be noted that the concentration of liver microsomal phospholipids can be increased with Zn deficiency in rats. Thus, the increased concentration of peroxidizable fatty acids in phospholipids may provide increased substrate for peroxidation (44).

A characteristic sign of Zn deficiency is flaking skin seborrhoea, a skin disorder suggested to be caused by abnormal skin cell membranes. Whereas the mechanism by which Zn deficiency induces this pathology is unknown, studies have demonstrated that peroxidation is higher than normal in skin cell membranes from Zn-deficient chicks, and that this effect is reduced with vitamin E supplementation (7).

Consistent with increased lipid peroxidation, a number of reports have demonstrated abnormal membrane morphology in tissues from Zn-deficient animals. Severe deterioration of cell membranes, as assessed by electron microscopy, was apparent in 11-d Zn-deficient fetuses (28). This time period corresponds to the time in which cell death occurs in the neural tube of Zn-deficient fetuses. The increased lipid peroxidation with resulting membrane damage during critical time periods may contribute to abnormal development in Zn-deficient fetuses (28,35).

Using a test system comprised of erythrocyte ghosts incubated in xanthine, xanthine oxidase, and Fe to generate oxygen radicals, Thomas et al. (47) demonstrated that Zn-mt provides effective protection against lipid peroxidation. Whereas some have suggested that mt functions as an antioxidant via interaction of thiol groups with oxygen radicals, the predominant role of this protein as an antioxidant appears to be its donation of Zn for incorporation into the membrane. Following its oxidation, thiol components of mt release the metal and allow Zn to compete with Fe for membrane binding sites, thereby reducing or preventing Fe redox-catalyzed lipid peroxidation. Whereas an actual physical role for mt-donated Zn as a means to prevent excessive lipid peroxidation in vivo has not been demonstrated, the concept is consistent with observations of increased tissue mt concentrations under similar conditions of oxidative stress (38).

Similar to Cu, the role of Zn as an antioxidant has been studied in situations where the animal is subjected to environmental or dietary conditions that increase the oxidative stress imposed on the animal. A few of these studies are discussed below.

Based on observations that serum and hepatic Zn concentrations are commonly low in cirrhotic patients, that lipid peroxidation is one mechanism by which cell necrosis develops in alcoholics, and that Zn functions as an antioxidant, the idea that Zn deficiency may augment ethanol-induced peroxidative damage has been tested by Sullivan et al. (44) and Dreosti et al. (21), using rats as models. However, using diene conjugation as an estimate of in vivo lipid peroxidation and TBARS as indicators of lipid peroxidation potential (in vitro peroxidation), Zn deficiency was not found to exacerbate the peroxidative effects of ethanol. Although these indices of lipid peroxidation were higher in the ethanol-treated Zn-deficient rats relative to Zn-adequate rats, the increase was as a result of the Zn deficiency independent of ethanol exposure.

Whereas the above studies were done in adult rats, Dreosti et al. (21) also investigated the potential interaction between Zn and ethanol on lipid peroxidation in the fetus. In this study, Zn deficiency alone resulted in a slight increase in TBARS in fetal liver microsomes; Zn deficiency plus ethanol resulted in values that were three times that observed in controls + ethanol, and two times that observed in Zn-deficient fetuses. Hence, the contribution of Zn and ethanol was more damaging to the fetal liver than either treatment alone.

One of the most extensively studied model systems of lipid peroxidation has been the hepatotoxicity produced by carbon tetrachloride. Consistent with a role for Zn as an antioxidant are results demonstrating that Zn^{+2} protects liver microsomes against lipid peroxidation damage induced by carbon tetrachloride metabolism (*12*).

Whereas it is common in any discussion on lipid peroxidation to focus on its negative consequences and resulting pathologies, it is important to bear in mind that abnormally low levels of lipid peroxidation may also be associated with negative outcome. For example, mitochondria and microsomes isolated from Morris hepatoma and Ehrlich ascites tumors exhibit lower levels of lipid peroxidation in vitro than control liver preparations, and the peroxidative capacity is inversely correlated to the growth rate of the tumor. Based on this, the effect of Zn deficiency on the capacity of liver and tumor mitochondrial and microsomal membranes to undergo lipid peroxidation has been assessed by Burke and Fenton (*9*). Analysis of conjugated dienes and TBARS demonstrated that lipid peroxidation was increased in tumor preparations from Zn-deficient rats relative to their pair-fed controls. Whereas it is tempting to interpret these results to suggest that tumors from Zn-deficient animals may then exhibit lower growth rates because of increased lipid peroxidation, these results may simply be reflective of Zn-induced alterations in fatty acid synthesis. Increases in phospholipid concentrations as well as alterations in membrane fatty acid composition in Zn-deficient animals, have been reported previously. An intriguing possibility is that membrane phospholipid concentrations and/or fatty acid composition may be a determinant of growth rate in some tumors. As such, the role of nutritional factors, such as Zn and Cu, whose levels of intake are associated with changes in membrane lipid composition, provides an exciting area of research into the factors regulating tumor growth.

4. MANGANESE

The concentration of Mn in animal tissue is normally less than 50 μM in soft tissues, and less than 0.5 nM in blood. The characteristic oxidation state of Mn in solution, in metal enzyme complexes, and in metalloenzymes, is Mn^{2+}. Similar to Fe^{3+}, Mn^{2+} has a high affinity for imidazole, in contrast to other divalent cations like Zn^{2+} and Cu^{2-} that have higher affinities for thiol.

Although considered rare, an experimental Mn deficiency results in a number of structural and physiological defects, including membrane damage, altered lipoprotein synthesis, bone defects, and abnormal carbohydrate metabolism (*36*). Defects that may be linked to a compromised antioxidant defense system are discussed below.

4.1. The Role of Mn in the Antioxidant Defense System

Manganese is an essential component of the enzyme Mn-superoxide dismutase (MnSOD), that is found primarily in the mitochondria (*25*). Manganese is present in this enzyme in the trivalent state; its catalytic role involves reduction, and then, reoxidation of the metal center during successive encounters with oxygen.

In addition to MnSOD, nonenzyme forms of Mn can play a role in the destruction, as well as the formation, of oxygen radicals. Low mol wt Mn complexes inhibit lipid peroxidation in in vitro systems containing microsomes, lysozomes, and cell membranes (*24*). Consistent with the in vitro data, although devoid of MnSOD, many lactic acid bacteria have a high tolerance for oxygen as a result of the ability of these organisms to accumulate high concentrations of Mn in the form of Mn polyphosphate that can dismutate O_2^{-} (*2*). Recently, Mn desferrioxamine complexes have also been shown to have SOD activity that can protect mammalian cells in culture from O_2^{-} damage induced by paraquat (*15*).

To investigate the mechanism by which Mn complexes act to reduce oxygen radical-induced damage, Cheton and Archibald (*10*) have utilized two different $OH^{\cdot}$-generating systems to assess the efficacy of Mn complexes as $OH^{\cdot}$ scavengers. When Mn^{2+}-pyrophosphate and Mn_2-polyphosphate, which are present in living cells (*10*), were added to incubation systems that generated $OH^{\cdot}$ via steps dependent on O_2^{-} and H_2O_2, the formation of $OH^{\cdot}$ was inhibited. In contrast, there was no effect on $OH^{\cdot}$-mediated damage when the same complexes were added to systems that generated $OH^{\cdot}$ independent of O_2 and H_2O_2. Based on these results, it can be suggested that Mn complexes act as antioxidants via their ability to reduce or block the formation of $OH^{\cdot}$, possibly through scavenging O_2^{-} or H_2O_2. Thus, whereas the ability of Mn to react with oxygen species is shared with that of Fe and Cu, there are fundamental differences between Mn and these two metals with respect to the mechanisms by which they react with oxygen complexes.

Similar to Cu,ZnSOD, the activity of MnSOD is affected by dietary status as well as environmental factors. A dietary deficiency of Mn can result in a reduction of MnSOD activity in rats, mice, and chickens (Table 3). In adult mice and rats fed Mn-deficient diets (1 μg Mn/g) prenatally and postnatally, the activity of this enzyme was significantly lower in liver, brain, heart, and lung than in tissues of animals fed control diets (45 μg Mn/g) (*18*). Similarly, Paynter (*39*) has reported a reduction in heart and kidney MnSOD activity in adult rats fed Mn-deficient diets from weaning on; however, in contrast to findings by deRosa et al. (*18*), liver MnSOD activity was only slightly lower in the deficient relative to control animals. This discrepancy, with regard to liver MnSOD activity, may be related to the differences in the duration and severity of the imposed Mn deficiency. In chickens, there was a depressed activity of MnSOD in liver after only 7 d of feeding a Mn-deficient diet to hatchlings (18). Concomitant with the decline in activity of MnSOD, in the chicken model, the activity of Cu,ZnSOD was increased, suggesting a compensatory response to the low MnSOD activity. In the chicken model, the activity of MnSOD is quickly elevated to normal by the reintroduction of Mn into the diet.

4.2. Functional Significance of Mn Deficiency-Induced Changes in the Free Radical Defense System

To assess the functional significance of lower than normal activity of MnSOD, Zidenberg-Cherr et al. (50) measured hepatic lipid peroxidation and MnSOD activity in Mn-sufficient and Mn-deficient rats from birth through sexual maturity. The activity of liver MnSOD increased from birth through 60 d of age in both groups. By d 60, MnSOD activity in Mn-deficient rats was half that observed in Mn-sufficient rats. Mitochondrial lipid peroxidation in Mn-deficient rats, as assessed by TBARS, was three times that observed in Mn-sufficient rats (Table 3). These findings suggest that the damage to mitochondrial membranes, previously observed in adult Mn-deficient animals (5), might be related in part to depressed MnSOD activity, which results in increased membrane lipid peroxidation. However, when mitochondrial ultrastructure was examined in Mn-deficient rats from birth through 60 d of age, no morphological damage was observed (52). In contrast, by 9 mo of age, liver from Mn-deficient rats showed abnormal mitochondria displaying large vacuoles in the matrix and disruptions in the outer double membrane (52). Whereas the finding of normal mitochondrial membranes in the young Mn-deficient animals argues against functional lipid peroxidation damage during this time period, damage in the older rats may be the result of excessive mitochondrial lipid peroxidation occurring in an earlier stage of development, as well as in later life. Similar to what may occur in Cu deficiency, increased peroxidation reactions may promote changes in the pattern of fatty acid synthesis such that the ratio of polyunsaturated to saturated fatty acid is altered. If such processes occur in the mitochondria, abnormal morphology of this organelle may develop. In addition, since Mn is a cofactor for several enzymes functioning in cholesterol and fatty acid synthesis (36), these processes may be affected by Mn deficiency, and thus, contribute to the formation of abnormal membranes.

In addition to diet, environment has also been shown to influence the activity of MnSOD. Activity of MnSOD can be induced under conditions that can result in an increased production of superoxide radicals, such as exposure to hyperbaric oxygen and ozone. Hyperoxia induces both MnSOD and catalase activity in pulmonary macrophages, whether the cells are incubated in vitro or if the animals are exposed in vivo (43). Similarly, ozone inhalation has been shown to increase total lung MnSOD and Cu,ZnSOD activity in mice. When Mn-deficient mice were

Table 3
Effects of Mn Deficiency on the Antioxidant Defense System in Liver

Species	Sex	Diet	Duration	MnSOD	GSH-Px	Catalase	GSH	LPO[1]	Ref
		(Mn ppm)							
Rat	M	0.2	1 mo	↓	ND[2]	–[3]	ND	↑	39
Mouse	M/F	1.0	Gest, 21 d	↓	–	–	–	–	18
Chicken	M/F	1.0	21 d	↓	ND	–	–	–	18
Rat	M/F	1.0	Gest, 60 d	↓	–	–	–	↑	50

[1]Includes measurements of TBARS, lipid hydroperoxides, and hydrocarbon gases as indices of lipid peroxidation.
[2]Not measured.
[3]No differences.

exposed to ozone, there was a decrease in MnSOD activity per gram lung and an increase in Cu,ZnSOD on a total lung basis (22). Thus, similar to the effect of dietary Cu deficiency on Cu,ZnSOD activity, the increase in MnSOD activity in response to ozone exposure is impaired by dietary Mn deficiency.

Similar to ozone and hyperbaric oxygen, ethanol exposure can (58) result in an increase in the activity of MnSOD. In rats (19,20), pigs (57), and primates (34), chronic ethanol consumption can result in increased MnSOD activity. That this increase in MnSOD activity is of benefit to the animal is suggested by the observation that Mn-deficient rats fed ethanol in their drinking water (20%) respond with extreme lethargy and anorexia (51). The mechanism by which agents such as, oxygen, ozone, and alcohol influence the biosynthesis and/or activity of these enzymes has not been determined.

Manganese deficiency can also influence an animal's response to ADR, the chemotherapeutic drug discussed above. When ADR is given to mice deficient in both Mn and vitamin E, there are marked elevations in mitochondrial lipid peroxidation as assessed by TBARS, compared to mice deficient in only Mn or vitamin E (53). Similarly, ultrastructural examination revealed that ADR-induced mitochondrial abnormalities in cardiac tissue were more severe in the low E, Mn-deficient mice than in mice fed diets deficient in only one of these nutrients (54). This observation further demonstrates the potential for synergism between components of the free radical defense system.

5. CONCLUSIONS

Taken together, information available supports a prominent role for Cu, Zn, and Mn nutriture in cellular defense from free radical damage. As one examines the literature, it becomes apparent that there are a number of inconsistencies with respect to the effects of the above mineral deficiencies on tissue lipid peroxidation. Since the methods most commonly used to assess lipid peroxidation (TBARS, diene conjugation, expirations of hydrocarbon gases) are dependent on fatty acid composition, changes in this parameter may influence the values reported for lipid peroxidation. For example, since Cu deficiency results in membranes that contain a higher ratio of saturated to polyunsaturated fatty acids, the "peroxidizability" of these membranes will be lower than that from membranes with a "normal" saturated to polyunsaturated ratio. Whereas at first glance, low lipid peroxidation values may appear contradictory to the hypothesis that Cu deficiency is associated with increased peroxidation, it must be emphasized that such alterations in the pattern of fatty acid synthesis may be a response to the increased free radical stress imposed on the cell as a result of the deficiency. The above issue illustrates some of the difficulties in interpreting the influence of trace element deficiencies on the antioxidant defense system. However, given the observations that mineral-induced changes in the free radical defense system can be functionally significant, this is an area of research that clearly merits further investigation.

REFERENCES

1. K. G. D. Allen, J. R. Arthur, P. C. Morrice, F. Nicol, and C. F. Mills. Copper deficiency and tissue glutathione concentration in the rat. *Proc. Soc. Exp. Biol. Med.* **187**, 38–43 (1988).
2. F. Archibald and I. Fridovich. The scavenging of superoxide radical by manganous complexes in vitro. *Arch. Biochem. Biophys.* **214**, 452–463 (1982).
3. P. S. Balevska, E. M. Russanov, and T. A. Kassabova. Studies on lipid peroxidation in rat liver by copper deficiency. *Int. J. Biochem.* **13**, 489–493 (1981).
4. G. M. Baroli, B. Giannattasio, P. Palozza, and A. Cittadini. Superoxide dismutase depletion and lipid peroxidation in rat liver microsomal membranes. *Biochim. Biophys. Acta* **966**, 214–221 (1988).
5. L. T. Bell and L. S. Hurley. Ultrastructural effects of manganese deficiency in liver, heart, kidney, and pancreas of mice. *Lab. Invest.* 29, 723–736 (1973).
6. W. J. Bettger, T. J. Fish, and B. L. O'Dell. Effects of copper and zinc status of rats on erythrocyte stability and superoxide dismutase activity. *Proc. Soc. Exp. Biol. Med.* **158**, 279–282 (1978).
7. W. J. Bettger, J. E. Savage, and B. L. O'Dell. A critical physiological role of zinc in the structure and function of biomembranes. *Life Sci.* **28**, 1425–1438 (1981).
8. T. M. Bray, S. Kubow, and W. J. Bettger. Effect of dietary zinc on endogenous free radical production in rat lung microsomes. *J. Nutr.* **116**, 1054–1060 (1986).
9. J. P. Burke and M. R. Fenton. Effect of zinc-deficient diet on lipid peroxidation in liver and tumor subcellular membranes. *Proc. Soc. Exp. Biol. Med.* **179**, 187–191 (1985).
10. P. L. B. Cheton and F. S. Archibald. Manganese complexes and the generation and scavenging of hydroxyl radicals. *Free Rad. Biol. Med.* **5**, 325–333 (1988).
11. K. Chung, N. Romero, D. Tinker, C. L. Keen, K. Amemiya, and R. B. Rucker. Role of copper in the regulation and accumulation of superoxide dismutase and metallothionein in rat liver. *J. Nutr.* **118**, 859–864 (1988).
12. M. Chvapil, J. N. Ryan, S. L. Elias, and Y. M. Peng. Protective effect of zinc on carbon tetrachloride-induced injury in rats. *Exp. Mol. Pathol.* **19**, 186–196 (1973).
13. M. Chvapil and C. F. Zukowski. New concept on the mechanism(s) on the biological effect of zinc. In *Clinical Applications of Zinc Metabolism*, W. J. Pories, W. H. Strain, J. M. Hsu, and R. L. Woosley, eds., C. Thomas, Springfield, IL, 1974, pp. 75–86.
14. M. S. Clegg, C. L. Keen, and L. S. Hurley. Biochemical pathologies of zinc deficiency. In *Zinc in Human Biology*, C. F. Mills, ed., Springer-Verlag, New York, 1988, pp. 129–145.
15. D. D. Darr, S. Yanni, and S. R. Pinnell. Protection of chinese hamster ovary cells from paraquat-mediated cytotoxicity by a low molecular weight mimic of superoxide dismutase (DF-Mn). *Free Rad. Biol. Med.* **4**, 357–363 (1988).

16. G. K. Davis. Microelement interactions of zinc, copper, and iron in mammalian species. *Ann. N. Y. Acad. Sci.* **355**, 98–108 (1980).
17. G. K. Davis and W. Mertz. Copper. In *Trace Elements in Human and Animal Nutrition*, vol. 1, W. Mertz, ed., Academic, London, 1987, pp. 301–364.
18. G. deRosa, C. L. Keen, R. M. Leach, L. S. Hurley. Regulation of superoxide dismutase by dietary manganese. *J. Nutr.* **110**, 795–804 (1980).
19. I. E. Dreosti and I. R. Record. Superoxide dismutase, zinc status, and ethanol consumption in maternal and fetal rat livers. *Br. J. Nutr.* **41**, 399–402 (1979).
20. E. I. Dreosti, I. R. Record, R. A. Buckley, S. J. Manuel, and F. J. Fraser. Ethanol and hepatic superoxide dismutase in rats. In *Trace Element Metabolism in Man and Animals (TEMA-4)*, J. McC. Howell, J. M. Gawthorne, C. L. White, eds., Griffin Press, Netley, South Australia, 1984, pp. 617–620.
21. I. E. Dreosti, I. R. Record, and S. J. Manuel. Zinc deficiency and the developing embryo. *Biol. Trace Element Res.* **7**, 103–122 (1985).
22. M. A. Dubick, S. Zidenberg-Cherr, R. Rucker, and C. L. Keen. Superoxide dismutase activity in lung from copper- and manganese-deficient mice exposed to ozone. *Toxicol. Lett.* **42**, 149–157 (1988).
23. M. Fields, R. J. Ferretti, J. C. Smith, and S. Reiser. Interaction between dietary carbohydrate and copper nutriture on lipid peroxidation in rat tissues. *Biol. Trace Element Res.* **6**, 379–391 (1984).
24. K. L. Fong, P. B. McCay, J. L. Poyer, B. B. Keele, and H. Misra. Evidence that peroxidation of lysosomal membranes is initiated by hydroxyl free radicals produced during enzyme activity. *J. Biol. Chem.* **248**, 7792–7797 (1973).
25. I. Fridovich. Superoxide dismutases. *Ann. Rev. Biochem.* **44**, 147–159 (1975).
26. I. M. Goldstein, H. B. Kaplan, H. S. Edelson, and G. Weissman. Ceruloplasmin: a scavenger of superoxide anion radicals. *J. Biol. Chem.* **254**, 4040–4045 (1979).
27. J. Harata, M. Nagata, E. Sasaki, I. Iishiguro, Y. Ohta, and Y. Marakami. Effect of prolonged alcohol administration on activities of various enzymes scavenging activated oxygen radicals and lipid peroxide levels in the liver of rats. *Biochem. Pharmacol.* **32**, 1795–1798 (1983).
28. A. J. Harding, I. E. Dreosti, and R. S. Tulsi. Zinc deficiency in the 11th day rat embryo: a scanning and transmission electron microscope study. *Life Sci.* **42**, 889–896 (1987).
29. K. E. Hill and R. F. Burk. Effect of selenium deficiency on the disposition of plasma glutathione. *Arch. Biochem. Biophys.* **240**, 166–171 (1985).
30. J. M. Hsu, W. L. Anthony, and P. J. Buchanan. Incorporation of glycine-1-14C into liver glutathione in zinc deficient rats. *Proc. Soc. Exp. Biol. Med.* **127**, 1048–1051 (1968).
31. S. K. Jain and D. M. Williams. Copper deficiency anemia: altered blood cell lipids and viscosity in rats. *Am. J. Clin. Nutr.* **48**, 637–64 (1988).
32. M. Jay, S. M. Stuart, C. J. McClain, D. A. Palmieri, and D. A. Butterfield. Alterations in lipid membrane fluidity and the physical state of cell-surface sialic acid in zinc-deficient rat erythrocyte ghosts. *Biochim. Biophys. Acta* **897**, 507–511 (1987).
33. S. G. Jenkinson, R. A. Lawrence, R. F. Burk, and D. M. Williams. Effects of copper deficiency on the activity of the selenoenzyme glutathione peroxidase, and one excretion and tissue retention of $^{75}SeO_3$. *J. Nutr.* **112**, 197–204 (1982).

34. C. L. Keen, T. Tamura, B. Lonnerdal, L. S. Hurley, and C. H. Halsted. Changes in hepatic superoxide dismutase activity in alcoholic monkeys. *Am. J. Clin. Nutr.* **41,** 929–932 (1985).

35. C. L. Keen and L. S. Hurley. Zinc and reproduction: effects of deficiency on fetal and postnatal development. In *Zinc in Human Biology,* C. F. Mills, ed., Springer-Verlag, New York, 1988, pp. 183–220.

36. C. L. Keen and T. W. Graham. Trace elements. In *Clinical Biochemistry of Domestic Animals,* 4th Ed., J. J. Kaneko, ed., Academic, New York, 1989, pp. 753–795.

37. R. A. Lawrence and S. G. Jenkinson. Effects of copper deficiency on carbon tetrachloride-induced lipid peroxidation. *J. Lab. Clin. Med.* **109,** 134–140 (1987).

38. S. H. Oh, J. T. Deagen, P. D. Whanger, and P. H. Weswig. Biological function of metallothionein. V. Its induction in rats by various stresses. *Am. J. Physiol.* **234,** E282–E285 (1978).

39. D. I. Paynter. The role of dietary copper, manganese, selenium, and vitamin E in lipid peroxidation in tissues of the rat. *Biol. Trace Element Res.* **2,** 121–135 (1980).

40. J. R. Prohaska and D. E. Gutsch. Development of glutathione peroxidase activity during dietary and genetic copper deficiency. *Biol. Trace Element Res.* **5,** 35–45 (1983).

41. N. Shiraishi, K. Aono, and K. Utsumi. Increased metallothionein content in rat liver induced by χ irradiation and exposure to high oxygen tension. *Radiat. Res.* **95,** 298–302 (1983).

42. T. H. Spence, S. G. Jenkinson, K. H. Johnson, F. J. Collins and R. A. Lawrence. Effects of bacterial endotoxin on protecting copper-deficient rats from hyperoxia. *J. Appl. Physiol.* **61,** 982–987 (1986).

43. J. B. Stevens and A. P. Autor. Proposed mechanism for neonatal rat tolerance to normobaric hyperoxia. *Fed. Proc.* **39,** 3138–3143 (1980).

44. J. F. Sullivan, M. M. Jetton, K. J. Hahn and R. E. Burch. Enhanced lipid peroxidation in liver microsomes of zinc-deficient rats. *Am. J. Clin. Nutr.* **33,** 51–56 (1980).

45. O. G. Taylor, W. J. Bettger, and T. M. Bray. Effect of dietary zinc or copper deficiency on the primary free radical defense system in rats. *J. Nutr.* **118,** 613–621 (1988).

46. P. J. Thornalley and M. Vasak. Possible role for metallothionein in protection against radiation-induced oxidative stress. Kinetics and mechanism of its reaction with superoxide and hydroxyl radicals. *Biochim. Biophys. Acta* **827,** 36–44 (1985).

47. J. P. Thomas, G. J. Bachowski, and A. W. Girotti. Inhibition of cell membrane lipid peroxidation by cadmium- and zinc-metallothioneins. *Biochim. Biophys. Acta* **884,** 448–461 (1986).

48. F. Ursini and A. Bindoli. The role of selenium peroxidases in the protection against oxidative damage of membranes. *Chemistry and Physics of Lipids* **44,** 255–276 (1987).

49. P. D. Van Helden and I. J. F. Wiid. Effects of adriamycin on heart and skeletal muscle chromatin. *Biochem. Pharmacol.* **31,** 973–977 (1982).

50. S. Zidenberg-Cherr, C. L. Keen, B. Lonnerdal, and L. S. Hurley. Superoxide dismutase activity and lipid peroxidation in the rat: developmental correlations affected by manganese deficiency. *J. Nutr.* **113,** 2498–2504 (1983).

51. S. Zidenberg-Cherr, L. S. Hurley, B. Lonnerdal, and C. L. Keen. Manganese deficiency: effects on susceptibility to ethanol in rats. *J. Nutr.* **115,** 460–467 (1985).

52. S. Zidenberg-Cherr, C. L. Keen, and L. S. Hurley. The effects of manganese deficiency during prenatal and postnatal development on mitochondrial structure and function in the rat. *Biol. Trace Element Res.* **7**, 31–48 (1985).
53. S. Zidenberg-Cherr and C. L. Keen. Influence of dietary manganese and vitamin E on adriamycin toxicity in mice. *Toxicol. Lett.* **30**, 79–87 (1986).
54. S. Zidenberg-Cherr and C. L. Keen. Enhanced tissue lipid peroxidation. Mechanism underlying pathologies associated with dietary manganese deficiency. In *Nutritional Bioavailability of Manganese*, C. Kies, ed., American Chemical Society, Washington, DC, 1987, pp. 56–66.
55. S. Zidenberg-Cherr, P. A. Benak, L. S. Hurley, and C. L. Keen. Altered mineral metabolism: a mechanism underlying the fetal alcohol syndrome in rats. *Drug-Nutrient Interact.* **5**, 257–274 (1988).
56. S. Zidenberg-Cherr, D. Dreith and C. L. Keen. Copper status and adriamycin treatment effects on antioxidant status in mice. *Toxicol. Lett.* **48**, 201–212 (1989).
57. S. Zidenberg-Cherr, C. H. Halsted, K. L. Olin, A. M. Reisenauer, and C. L. Keen. The effect of chronic alcohol ingestion on free radical defense in the miniature pig. *J. Nutr.*, **120**, 213–217 (1990).
58. S. Zidenberg-Cherr, K. L. Olin, J. Villanueva, A. Tang, S. D. Phinney, C. H. Halsted, and C. L. Keen. Ethanol-induced changes in hepatic free radical defense mechanisms and fatty acid composition in the miniature pig. *Hepatology*. In press.

From: *Trace Elements, Micronutrients, and Free Radicals* • Ed.: I. E. Dreosti • ©1991 The Humana Press Inc.

CHAPTER 6

Vitamins
and Related Dietary Antioxidants

CHING K. CHOW

ABSTRACT

Reactive oxygen species such as superoxide radicals, hydroxyl radicals, hydrogen peroxide, and lipid hydroperoxide are injurious to cellular constituents. Normally, the metabolic machinery of the cell is able to prevent or revert most of the adverse effects of oxidative stress. Vitamins and related compounds may directly or indirectly be involved in one or more stages of cellular antioxidant defense. They may interact with oxidants or oxidizing agents directly (e.g., ascorbic acid), scavenge free radicals and singlet oxygen (e.g., vitamin E, ascorbic acid and beta-carotene), be involved in the removal or separation of transition metals from the specific site of action by chelators or via membrane barrier, or participate in the repair or replacement of damaged molecules and cells by biosynthetic and other processes (e.g., vitamin A and B vitamins). Certain vitamins can act synergistically, or antagonistically, according to the levels or activities of other antioxidant defense systems involved. Many vitamins are interrelated and appear to act compensatorily or complimentarily with the others in different stages of the overall antioxidant defense.

1. INTRODUCTION

Whereas oxygen is essential for life, it may also be harmful. The adverse effects resulting from inhalation of high concentration of inspired oxygen have long been recognized. Since the reactivity of oxygen was too slow and too limited to account for the rate at which toxic effects developed, most of the damaging effects of oxygen have been attributed to the formation of more reactive oxygen radicals rather than the ground state of molecular oxygen *per se* (*16,25,26*).

While subjected to various types of oxidative stress, such as that caused by inhaled oxidants or ingested oxidizing chemicals, the metabolic activity of the cell is able to control or prevent adverse oxidative reactions under normal conditions. However, when the antioxidant potential is weakened or oxidative stress is greatly increased, irreversible damage to the cell may occur. The susceptibility of a given organ or organ system to oxidative damage can be described as a function of the overall balance between the factors that exert oxidative stress and those that exhibit antioxidant potential. In other words, oxidative damage may be regarded as a consequence of insufficient antioxidant potential.

Recently, the possible role of free radical-induced oxidative tissue injury in the toxicity of chemicals, drugs, and environmental agents, as well as in the pathogenesis of certain degenerative diseases has received considerable research attention. From the substantial knowledge accumulated during the past decade, this chapter will deal mainly with the possible role of vitamins and related dietary antioxidants in cellular defense against oxidative damage.

2. FREE RADICAL-INDUCED TISSUE DAMAGE

Oxygen generally needs to be activated to participate in biological redox reactions. Reactive oxygen species have long been recognized as intermediates of many essential biological redox reactions. Phagocytosis and bactericidal activity of polymorphonuclear leukocytes and alveolar macrophages, for example, are normally accomplished by a burst of active metabolism and the generation of reactive oxygen species (*19,30*). The adverse effects induced by reactive oxygen species generated within or adjacent to the cellular environment have been suggested to be partly responsible for the tissue injury resulting from exposure to a large variety of chemicals, drugs, and environmental agents. The oxidative damage mechanism has also been implicated in the pathogenesis of certain degenerative diseases (*16,17,21,25*).

A number of activated oxygen species detected in living systems have been shown to be capable of reacting with and/or inactivating essential biological materials. They may subsequently be converted into another compound, or degraded. The secondary products of these reactive oxygen species may or may not be more reactive or harmful to biological systems. Among the reactive oxygen intermediates, lipid hydroperoxides, hydroxyl radicals, superoxide radicals, and hydrogen peroxide are considered to be the most biologically significant.

2.1. Lipid Hydroperoxides

In the presence of a free radical or a free radical initiator, biological materials, particularly cell membranes that contain relatively high proportions of polyunsaturated lipids, become susceptible to oxidation. The process, known as lipid peroxidation, has been associated with the loss of polyunsaturated fatty acids, and the formation of lipid hydroperoxides. Whereas lipid hydroperoxides are injurious to the cells, they may be detoxified/metabolized by glutathione peroxidase systems (32,45). However, in the presence of some transition metals, lipid hydroperoxides may also be cleaved homolytically to form more free radicals and, thus, accelerate peroxidation of membrane lipids. Furthermore, lipid hydroperoxides may also be degraded to form aldehydes and other secondary products, and some of them are known to be cytotoxic (16,17,34).

If the process of free radical-induced chain reaction is not terminated and the peroxidized membrane not repaired, it may disturb the fine structure of biological membranes, and affect the permeability and functions of the membrane. Lipid hydroperoxides and other products, such as malonaldehyde, may react with and inactivate essential proteins, enzymes, and nucleic acids. If the damage to essential cellular constituents is not repaired, the process may cause irreversible damage to the cells. Eventually, it may lead to the death or turnover of the cell (15–17,34).

2.2. Hydroxyl Radicals

The hydroxyl radical is very reactive. In addition to being capable of initiating lipid peroxidation via proton abstraction, hydroxyl radicals can exert other biological effects in aqueous solution via addition and electron transfer mechanisms (2,26,43). They can react readily with almost every type of molecule found in the living cell, e.g., sugar, amino acid, phospholipids, nucleotides, and organic acids. Because of its high reactivity, hydroxyl radicals have been suggested to be the initiator required for many oxidative reactions, including lipid peroxidation in vivo. On the other hand, hydroxyl radicals may be too reactive to survive collisions with compounds adjacent to the site of formation, and travel to the critical target site. Thus, the site-specific Haber-Weiss reaction for the formation of hydroxyl radicals may be of critical importance in determining the reactivity/toxicity of hydroxyl radicals (27,43).

2.3. Superoxide Radicals

Superoxide radicals can be generated during the respiratory burst of phagocytic cells and many other biological redox reactions (19,21,29). Results obtained from in vitro studies suggest that the superoxide radical is not very reactive, especially in aqueous solution (*26,38*). Since hydroxyl radical scavengers are capable of protecting damage induced by the superoxide generation systems, hydroxyl radicals generated, possibly via the iron-catalyzed mechanism, rather than superoxide radicals *per se*, are likely the agent responsible for the damaging effect of superoxide radicals in the aqueous phase (*21,25,26,38*). Superoxide radicals, however, are capable of exerting deleterious effects independent of participating with hydrogen peroxide in the production of the hydroxyl radical. Superoxide radicals can attack and inactivate a number of essential macromolecules, including catalase and glutathione peroxidase, although more slowly than hydroxyl radicals.

2.4. Hydrogen Peroxide

Hydrogen peroxide can be produced during phagocytosis, via the enzymatic dismutating action of superoxide dismutase, and many other biological reactions involving molecular oxygen (*19,21,26*). Hydrogen peroxide is a weak oxidizing agent. However, it can inactivate sulfhydryl enzymes (*26*). Whereas the peroxide is not very reactive in the aqueous phase at physiological concentrations, it can cross biological membranes. Because of the possible involvement of hydrogen peroxide in the generation of hydroxyl radicals (*25,26*), this property places hydrogen peroxide in a more prominent role in initiating cytotoxicity than its chemical reactivity indicates. DNA damage, for example, has been related to increased hydrogen peroxide generation by hypolipidemic drug-induced peroxisomes (*18*), although hydrogen peroxide may not be the agent directly responsible.

3. VITAMINS AND RELATED COMPOUNDS IN CELLULAR ANTIOXIDANT DEFENSE

In view of the potential adverse effects of oxygen and its reactive intermediates, it is important that the cell possesses various mechanisms that are capable of protecting against the deleterious effects of reactive oxygen species under the cellular environment. Normally, the cell may control or prevent oxidative damage by physically separating oxygen and reactive intermediates from susceptible cellular components, by providing molecules that effectively compete for or inactivate the activated oxygen species, and by lysing, removing, or by repairing damaged molecules. All the antioxidant defense mechanisms involve the complex organization of intracellular components.

In addition to biomembranes that physically separate the cell into many compartments, many enzymic and nonenzymic systems may function as antioxidants in the cell. Based on the mode and stage of action, the antioxidant defense systems can be classified into the following five categories:

1. Direct interaction with oxidants and oxidizing agents by ascorbic acid, glutathione, and other "sacrificing" compounds.
2. Scavenging of free radicals and singlet oxygen by vitamin E, beta-carotene, superoxide dismutase, and other scavengers.
3. Separation or removal of transition metals from the specific site of action by cell membranes, and by binding proteins and other chelators.
4. Reduction or metabolism of hydroperoxides by glutathione peroxidase and catalase.
5. Replacement or repair of the damaged membranes and molecules by a variety of molecules and biosynthetic activities.

Nutrients are essential for fundamental cellular processes. Micronutrients, especially vitamins, play essential and unique roles in overall cellular antioxidant defense. Vitamins are defined as organic substances, needed in very small quantity, that perform specific functions, and must be provided in the diet. The possible role of vitamins and related compounds in modulating cellular antioxidant defense is summarized in Table 1. The following discussion briefly examines each vitamin in terms of its function and possible role in cellular antioxidant defense.

3.1. Lipid-Soluble Vitamins

3.1.1. Vitamin A and Carotenoids

The term vitamin A is used for beta-ionone derivatives, other than carotenoids, that have biological activity of all-trans retinol. Forms of vitamin A include retinol, retinal, and retinyl esters (37,40,47). Vitamin A is normally stored in the liver as fatty acyl esters, and can be mobilized and delivered to the tissues in the form of the retinol-binding protein complex.

Vitamin A is indispensable for vision, reproduction, maintenance of epithelial integrity, as well as normal cell growth and development (37,40,47). Thus, vitamin A may reduce oxidative stress by maintaining the integrity of epithelium, or alleviate oxidative damage by facilitating the replacement or repair of damaged molecules and cells. Vitamin A has also been shown to play a role in regulating the levels of ceruloplasmin (5), a copper containing acute phase plasma protein, and an important extracellular antioxidant (24).

Excessive vitamin A, however, may be associated with increased oxidative stress by enhancing vitamin E oxidation and/or turnover (39), and by decreasing the activities of superoxide dismutase and glutathione peroxidase (39,41). High levels of vitamin E, on the other hand, have been shown to spare vitamin A by increasing liver stores and reducing turnover (39), and to eliminate some toxic effects of vitamin A (42). In addition to vitamin E, interactions between vitamin A and other vitamins

Table 1
Possible Role of Vitamins and Related Compounds in Cellular Antioxidant Defense

Vitamins	Coenzyme	Possible mode of action or mechanisms
Vitamin A	—	Maintain epithelial integrity; regulate cell growth and proliferation; modulate ceruloplasmin levels
Carotenoids	—	Quench singlet oxygen and free radicals; serve as precursors of vitamin A
Vitamin D	—	Modulate redox status through calcium homeostasis
Vitamin E	—	Scavenge free radicals and singlet oxygen; stabilize cell membranes; supress xanthine oxidase synthesis
Vitamin K	—	Involved in electron-transport chain
Vitamin C	—	Directly interact with oxidants/oxidizing agents; scavenge free radicals and singlet oxygen; regulate iron absorption and metabolism; involved in the regeneration of vitamin E and conversion of folic acid to coenzyme form
Thiamin	Thiamin pyrophosphate	Involved in the synthesis of nucleic acids and conversion of tryptophan to niacin
Riboflavin	FMN, FAD	Participate in electron-transport chain (FMN); involved in the conversion of tryptophan to niacin, as well as folic acid and vitamin B_{12} to coenzyme forms; regulate glutathione reductase activity
Niacin	NADH, NADPH	Directly interact with oxidants/oxidizing agents; maintain glutathione, ascorbic acid, hemoglobin, and other compounds in reduced states; provide energy and reducing equivalents for metabolic activities; involved in the conversion of folic acid to coenzyme form

Pyridoxine	Pyridoxal phosphate	Involved in the synthesis of nucleic acids, elastin, and glutathione, in the conversion of tryptophan to niacin, and in the generation of energy
Folacin	Tetrahydrofolic acid	Involved in the synthesis of thymidine nucleotides of DNA, and in the formation of red blood cells
Cobalamin	—	Participate in the synthesis of thymidine nucleotides of DNA, and in the development of red blood cells
Pantothenic acid	Coenzyme A	Provide energy for reducing equivalents, and for synthetic activities
Biotin	—	Participate in energy generation and biosynthesis
Coenzyme Q	—	Participate in electron-transport chain; act as an antioxidant when in the reduced state
Choline	—	Involved in the synthesis of membrane phospholipids
Inositol	—	Involved in the synthesis of membrane phospholipids
Lipoic acid	—	Involved in the generation of energy and reducing equivalents
Para-aminobenzoic acid	—	Serve as a precursor of folic acid
Bioflavinoids	—	Enhance the utilization of ascorbic acid

have also been noted. High doses of vitamin A, for example, have been found to reduce tissue storage of ascorbic acid, and to manifest vitamin K deficiency. Excessive vitamin A intake, on the other hand, has been shown to protect against certain adverse effects of vitamin D toxicity in experimental animals (7,8).

Carotenoids are red and yellow fat soluble C-40 pigments with structures based on a tetraterpenoid skeleton, synthesized mainly by plants. Several carotenoids, especially beta-carotene, exhibit vitamin A activity. Approximately six parts of beta-carotene or 12 parts of other carotenoids are equivalent to the vitamin A activity of 1 part of retinol. Whereas increased plasma content has been observed, a large dose of carotenoids does not result in a significantly higher concentration of vitamin A, or cause hypervitaminosis A (46).

Beta-carotene has been shown to be an effective quencher of singlet oxygen, or function as an effective radical-trapping antioxidant (11,20). Activated beta-carotene can either give up its activation energy as heat, or absorb the heat internally and alter its molecular configuration. Beta-carotene may also react with singlet oxygen through covalent bonding and generate a ketone or peroxide.

Beta-carotene may exert other antioxidant functions in addition to the quenching of singlet oxygen. For example, it has been shown to inhibit lipid peroxidation initiated by xanthine oxidase (29). Prior treatment with beta-carotene has also been shown to reduce lipid peroxidation of guinea pigs injected with CCl_4, as measured by ethane and pentane production (31). However, beta-carotene does not seem to be a conventional antioxidant. Studies have shown that beta-carotene exhibits good free radical-trapping antioxidant behavior only at a partial pressure of oxygen significantly less than 150 torr, the pressure of oxygen in normal air (12). Such low oxygen partial pressures are found in most tissues under physiological conditions. At higher oxygen pressures, beta-carotene loses its antioxidant activity and shows an autocatalytic prooxidant effect, particularly at relatively high concentrations. (12).

3.1.2. Vitamin D

Vitamin D_2 (ergocalciferol) and vitamin D_3 (cholecalciferol) can be formed when their respective precursors, ergosterol (plant origin) and 7-dehydrocholesterol (animal origin), are exposed to sunlight (ultraviolet). Since the precursors of vitamin D can be produced in the body, it is considered by some to be a hormone rather than a vitamin. A major function of vitamin D is to regulate the absorption of dietary calcium and phosphorus and, thus, provide the optimal amounts of these two minerals needed for bone mineralization.

Vitamin D does not directly participate in the cellular redox reactions. As a result of its regulating role in calcium homeostasis, however, vitamin D may be indirectly involved in the control of oxidative damage. Calcium is a universal regulator of many metabolic processes. It has been shown that calcium ions appear to have a biphasic effect on free radical lipid peroxidation (3). The stimulating effect of lipid peroxidation by low calcium concentrations (10^{-6} M) seems to be related to its ability to release bound ferrous ions, and the inhibitory effect of high calcium concentrations may be caused by its interaction with superoxide radicals (3).

3.1.3. Vitamin E and its Quinones

Vitamin E is the best known fat-soluble antioxidant. The term "vitamin E" refers to at least eight toco and tocotrienol structures possessing vitamin E activity. Alpha-tocopherol is predominant in most species, and is significantly more potent than any other naturally occurring tocopherol. The primary role of vitamin E in preventing free radical-initiated lipid peroxidation damage has been accepted by most investigators in the field. Increasing evidence indicates that vitamin E may exert its biological function within the cellular membrane. The localization of vitamin E within the hydrophobic bilayer of cellular membrane as a complex with the polyunsaturated fatty acids of phospholipids (14,33) supports this view.

The protective effect of vitamin E against lipid peroxidation tissue damage has been attributed to its ability to scavenge free radicals and, thus, terminate the free radical chain reaction. In addition to its function as a free radical chain-breaker, the membrane localization property of vitamin E may serve to stabilize the membrane structure and regulate its functions (14,33). Vitamin E has also been shown to be capable of scavenging singlet oxygen (23). However, because of the extremely high reactivity of singlet oxygen, it is possible that this property of vitamin E may be secondary to its function as a free radical chain breaker.

It is not clear whether vitamin E is directly associated with the biosynthesis and functions of any specific enzyme. However, it has been shown that vitamin E deficiency produces a marked increase in liver xanthine oxidase activity in rabbits and rats by increasing its *de novo* synthesis (13). In view of the potential role of the xanthine oxidase system in generating superoxide radicals (21,29), it is possible that vitamin E may also protect the cell from oxidative injury by inhibiting the synthesis of xanthine oxidase. More studies are needed to substantiate this hypothesis.

Tocopheryl quinones are the most common oxidation products of tocopherols known (15). As alpha-tocopherol is the dominant form of vitamin E in humans and animals, it is not surprising that only alpha-tocopheryl quinone, but not other forms, has been detected in the tissue. Whereas alpha-tocopheryl quinone and its hydroquinone have been suggested to possess antioxidant property under certain conditions, it is unlikely to be of biological significance, owing to its small quantity and because of its effective metabolic pathway.

3.1.4. Vitamin K

The term vitamin K refers to a group of substances belonging to a family of chemical substance called quinones. Phylloquinone (vitamin K_1), synthesized by plants and menaquinone (vitamin K_2), synthesized by bacteria, are the major forms found in nature. The synthetic form of vitamin K, menadione, is more potent than either phylloquinone or menaquinone. Vitamin K is an essential factor for blood clotting in humans and animals. The quinone structure of vitamin K compounds enables them to act as an electron acceptor in oxidation–reduction reactions. Menadione, for example, has been shown to be involved in the formation of hydroxyl radicals via the Haber-Weiss reaction (36), and the oxidation of NAD(P)H, that results in the inhibition of aerobic glycolysis, depletion of the mitochondria ATP, and the loss of the flux of ionized calcium across the mitochondria and cellular membranes.

3.2. Water-Soluble Vitamins

3.2.1. Ascorbic Acid

L-Ascorbic acid (vitamin C) is a reducing agent and an important water-soluble vitamin for humans and certain species of animals. Ascorbic acid functions to maintain sulfhydryl compounds, including GSH, in a reduced state, and participates in many redox reactions. The vitamin is capable of scavenging free radicals and singlet oxygen, and may also directly interact with oxidants and oxidizing agents. Ascorbic acid at physiological concentrations can also scavenge the myeloperoxidase-derived oxidant hypochlorus acid at rates sufficient to protect important biological targets, such as alpha-1 antiprotease, against inactivation (28).

As an important extracellular antioxidant, ascorbic acid is functionally interrelated to both vitamin E and glutathione. Experimental evidence suggests that both ascorbic acid and glutathione may be involved in the regeneration of vitamin E, and glutathione and NADH are involved in maintaining ascorbic acid in a reduced state (15,35). Whereas glutathione is an effective co-antioxidant with vitamin E during peroxidation of liposomes, it is not effective with water-soluble antioxidant, 6-hydroxy-2,5,7,8-tetramethylchroman-2-carboxylate (Trolox), an analog of vitamin E (6). Since glutathione does not seem to act synergistically in regenerating vitamin E or Trolox, the mode of antioxidant action of glutathione is suggested to trap peroxyl radicals in the aqueous phase and, thereby, indirectly spare vitamin E in the bilayer (6). More studies are needed to clarify the role of ascorbic acid and glutathione in the sparing and regeneration of vitamin E in vivo.

Ascorbic acid has multiple effects on cellular redox systems. It has long been recognized that ascorbic acid is functionally related to iron metabolism. In addition to its enhancing effect on iron absorption, ascorbic acid can increase the stability of iron-binding proteins, including ferritin, intracellularly (*10*). Iron may facilitate the decomposition of lipid hydroperoxides, the formation of hydroxyl radical from hydrogen peroxide and superoxide, and the generation of superoxide radical and hydrogen peroxide, and other oxidative reactions (*25–27*).

It is generally accepted that the ferrous form, but not the ferric form, of iron has the ability to catalyze oxidative reactions. A key element of the cellular antioxidant defense is, therefore, to maintain transition metals, such as iron and copper bound to proteins and, thus, not available for catalytic reactions. The ability of ascorbic acid to enhance the release of transition metals from protein complexes, and to reduce them to catalytic forms, has implicated this compound to be a prooxidant as well. Ethanol-induced lipid peroxidation of biological biomembranes, for example, was potentiated by ascorbic acid (*1*). Higher concentration of sodium ascorbate in solution has also been shown to induce erythrocyte damage in premature infants (*4*). Also, evidence available indicates that large dose of ascorbic acid can cause erythrocyte damage in vivo (*44*). It appears that the concentrations and subcellular distribution are important in determining the antioxidant or prooxidant functions of ascorbic acid.

3.2.2. Thiamin (Vitamin B_1)

Thiamin, a sulfur- and amine-containing compound, occurs in a wide variety of plant and animals foods. The addition of two phosphate groups produces the primary physiologically active form, thiamin pyrophosphate. In this form, thiamin acts as a coenzyme (cocarboxylase) in energy generating reactions (glycolysis and the Krebs cycle) involving carbohydrates, fatty acids, and amino acids. Thiamin is also involved in the metabolism of nucleic acids. Thus, in addition to contributing to the generation of reducing equivalents, thiamin may play an important role in cellular antioxidant defense via the repair of the damaged molecules. Furthermore, thiamin contributes to the conversion of the amino acid, tryptophan, to another vitamin, niacin.

3.2.3. Riboflavin (Vitamin B_2)

Riboflavin occurs largely in dairy products and other animal foodstuff. When in its coenzyme form, it plays a vital role in the release of energy from carbohydrate, lipids, and proteins. The two coenzymes, flavin monophosphate (FMN) and flavin adenine diphosphate (FAD), are attached to a variety of proteins and are known as flavoproteins. The flavoproteins are a group of the carrier molecules to which hydrogen becomes attached as it moves through the electron-transport chain. Through its role in activating vitamin B_6, riboflavin is involved in the conversion of tryptophan to niacin. Also, riboflavin is needed for the conversion of folic acid to its coenzymes, and their subsequent storage in the body.

Since riboflavin is also involved in DNA synthesis, it is expected to have a direct effect on cell proliferation, as well as the ability of the cell to repair the damaged molecules. In addition, riboflavin may play a role in cellular antioxidant defense by regulating the activity of glutathione reductase (15).

3.2.4. Niacin

The term niacin includes nicotinic acid, nicotinamide, and related compounds. Approximately 60 parts of tryptophan can be converted to one part of niacin, and the conversion involves three other vitamins—thiamin, riboflavin, and pyridoxine. Niacin is involved in various metabolic reactions of the cell in a manner similar to riboflavin. The coenzymes of niacin, nicotinamide adenine dinucleotide (NAD), and nicotinamide adenine dinucleotide phosphate (NADP), exist in both oxidized and reduced forms, and take part in many of the hydrogen-transfer reactions necessary for the utilization of glucose, amino acids, and fatty acids. The reduced forms of these coenzymes (NADH and NADPH) may serve as antioxidants by directly reacting with oxidants or indirectly by maintaining other antioxidants, such as ascorbic acid and glutathione, in a reduced state (15,35). NADPH is also an essential substrate for glutathione reductase, which acts enzymatically to maintain glutathione in a reduced state (15).

3.2.5. Pyridoxine (Vitamin B_6)

Vitamin B_6 is a complex of three closely related compounds, pyridoxine, pyridoxal, and pyridoxamine. All three forms are functionally active and can be converted to the active coenzyme, pyridoxal phosphate. Pyridoxal phosphate is required in transamination, decarboxylation, transsulfuration, and side-chain transfers essential for the synthesis and breakdown of amino acids. Pyridoxine also plays a role in the synthesis of hemoglobin, elastin, and nucleic acids, and in the release of glycogen from the liver and muscle. Thus, pyridoxine may also participate in the overall cellular antioxidant defense by modulating substrates for the DE NOVO biosynthesis of glutathione, a tripeptide consisting of cysteine, glycine, and glutamic acid, and by providing effective repair systems. As mentioned previously, pyridoxine is also involved in the conversion of tryptophan to niacin.

3.2.6. Folacin

Folacin is a term that comprises folic acid and related compounds. Its coenzyme form, tetrahydrofolic acid, is necessary for single carbon (formyl and methyl) transfer during various reactions, including synthesis of the nucleotide base, thymine, of nucleic acids, and the formation of red blood cells. Thus, folacin may play a role in the overall antioxidant defense through its participation in the repair of damaged molecules or cells. Ascorbic acid and niacin are involved in the conversion of folic acid to tetrahydrofolic acid.

3.2.7. Cobalamin (Vitamin B₁₂)

Cobalamin is one of the most potent forms of vitamin B_{12}, a large complex molecule that contains cobalt. Vitamin B_{12} participates in some enzymic reactions involving the transfer of methyl groups, and the shift of a hydrogen atom from one carbon atom to an adjacent one. Similar to folacin, it is necessary for the formation of the thymidine nucleotides of DNA, and the development of red blood cells. Vitamin B_{12}, along with biotin, also functions in the metabolism of odd-numbered fatty acids that are important in myelin formation. Thus, the vitamin may be involved in the replacement or repair of damaged molecules or cells caused by oxidative reactions.

3.2.8. Pantothenic Acid

Pantothenic acid is a relatively simple compound. It functions as a component of coenzyme A (CoA). As part of CoA, pantothenic acid participates in many different enzymic reactions involving not only acetyl groups but acyl groups in general. Because of its central role in energy metabolism, it is vital to all energy requiring processes within the body. Through its coenzyme form, pantothenic acid is involved in generating energy necessary for biosynthetic activity, and thus, plays an important role in repairing or replacing damaged molecules.

3.2.9. Biotin

Like thiamin, biotin is a sulfur-containing compound. It acts as a carbon dioxide carrier in carboxylation reaction that lengthen carbon chain. In addition to fatty acid synthesis, biotin is involved in several transcarboxylation reactions of amino acids, and in protein and carbohydrate metabolism. Whereas no specific antioxidant function is known for biotin, its participation in a variety of biological reactions suggest that this vitamin may be indirectly involved in the repair process similar to the other B vitamins.

3.3. Vitamin-Like Compounds

In addition to the abovementioned vitamins, there are other substances that have some properties of vitamins but fail to meet all the criteria necessary to be classified as vitamins. This group of compounds include coenzyme Q, choline, inositol, lipoic acid, para-aminobenzoic acid, and the bioflavinoids.

3.3.1. Coenzyme Q (Ubiquinone)

Coenzyme Q, also known as mitoquinone and ubiquinone, is a very versatile molecule synthesized in animal tissue and functionally involved in a number of distinct, but related, cellular processes. As an obligatory electron-transport chain component in mitochondria, coenzyme Q functions to provide a mobile link, allowing reversible interactions between NADH dehydrogenase, succinate dehydrogenase, and cytochrome-b-c_1 portions of the electron transport chain. Free radicals are produced during mitochondrial electron-transfer-chain activity, and the rate of superoxide radical formation is directly related to the rate of oxygen utilization. Coenzyme Q, when in the reduced state, on the other hand, may act as an antioxidant, protecting the components of mitochondrial membranes from peroxidative damage (9). Endurance training has been reported to increase the mitochondrial coenzyme Q, but not vitamin E, content in skeletal muscles, suggesting an increase in oxidative stress (22).

3.3.2. Choline

Choline can be synthesized in the body from glycine, providing a source of methyl groups and adequate amounts of folacin and cobalamin are available. Choline is a precursor of the important neurotransmitter, acetylcholine. It is also a component of the phospholipid, lecithin, which forms an essential part of cell membranes and lipoproteins. Deficiency of choline is associated with structural and functional abnormalities in cells. As a component of membrane phospholipids, choline is conceivably essential for membrane repair and replacement.

3.3.3. Inositol

The structure of inositol is similar to that of glucose, and it is found widely in foods. Its primary function appears to be as a component of phospholipids. Similar to that of choline, the possible role of inositol in antioxidant defense appears to be related to the structural integrity of biomembranes.

3.3.4. Lipoic Acid

Lipoic acid is a sulfur-containing compound synthesized in mammalian tissues, and is essential for the growth of several microorganisms. It acts as a coenzyme in conjunction with thiamin, niacin, riboflavin, and pantothenic acid in the conversion of pyruvate to acetyl CoA, and of alpha-ketoglutarate to succinyl CoA. As one of the important factors in energy metabolism, lipoic acid may also be involved indirectly in the repair process similar to thiamin, niacin, riboflavin, and pantothenic acid.

3.3.5. Para-Aminobenzoic Acid

Para-aminobeinzoic acid is a growth factor for bacteria and lower animals. It is a component of folic acid, and can satisfy the need in rats and mice for dietary folic acid. Thus, its possible involvement in antioxidant defense is likely to be associated with the action of folic acid. No other metabolic role of this compound has been established.

3.3.6. Bioflavinoids

The bioflavinoids refer to a mixture of biologically active phenolic compounds widely distributed in plants. Whereas the bioflavinoids have been associated with the enhancement of ascorbic acid utilization, no specific biologic role has been found.

4. INTERRELATIONSHIP OF VITAMINS AND RELATED COMPOUNDS IN ANTIOXIDANT DEFENSE

Various antioxidant defense systems are functionally interrelated, and appear to act in a concerted manner in overall antioxidant defense (15). Evidence available indicates that certain defense systems may act synergisitically or antagonistically, depending on their levels or activities involved. Some are able to compensate for others, or to respond adaptively under certain conditions. Furthermore, some appear to act complementarily and interdependently to achieve optimal functionality. Accordingly, it is not possible to understand the function of vitamins and related compounds in antioxidant defense without referring to their interactions with other defense systems. The possible involvement of vitamins and related compounds in overall antioxidant defense and their relationship with others are summarized in Fig. 1.

Owing to the prominent role of vitamin E and ascorbic acid in cellular antioxidant defense, the association of ascorbic acid with the regeneration of vitamin E has received considerable interest. In the process of quenching free radicals, vitamin E is first converted to the tocopheryl chromanyl radical. At this stage, the tocopherol radical can be further oxidized irreversibly to tocopheryl quinone or be reverted to tocopherol by a system or group of systems that are not yet completely defined, including ascorbic acid, glutathione, and possibly, specific tocopherol regenerating enzyme/protein (15,35).

An important aspect of the cellular antioxidant defense that has not received sufficient attention is the repair and/or replacement of damaged molecules and cells. Almost all B-vitamins are directly or indirectly involved in providing energy and reducing equivalents, and in the biosynthesis of essential molecules, including DNA. In addition to their specific functions, many B vitamins are closely interrelated. For example,

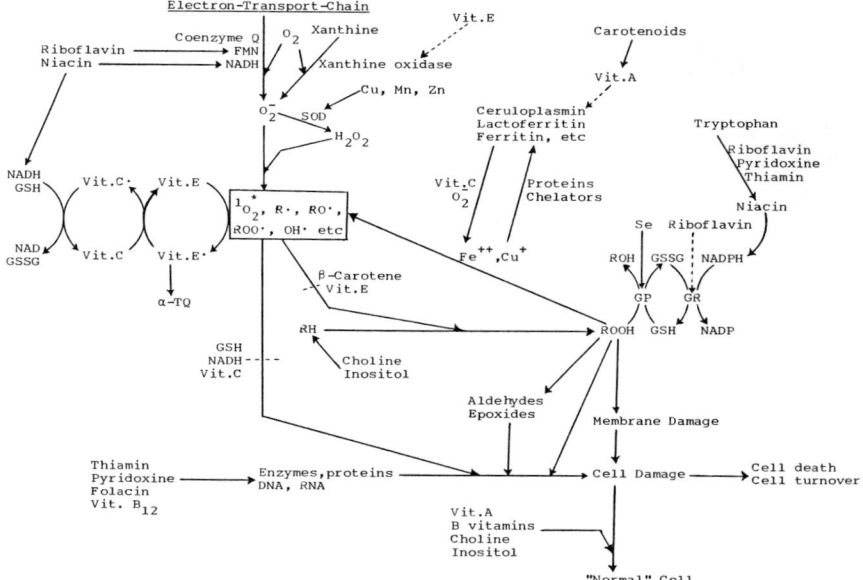

Fig. 1.   Possible role of vitamins and related compounds in cellular antioxidant defense and their interrelationship. RH represents membrane lipids; ROOH, peroxidized membrane lipids; GP, glutathione peroxidase; GR, glutathione reductase; GSH, reduced glutathione; GSSG, oxidized glutathione; SOD, superoxide dismutase; O_2, superoxide radical; O_2, singlet oxygen; OH˙, hydroxyl radical; R˙, RO˙, ROO˙, free radicals; vit.E˙, vitamin E radical; α-TQ, α-tocopheryl quinone; vit.C˙; semidehydro ascorbate; H_2O_2, hydrogen peroxide.

thiamin, riboflavin, and pyridoxine are all involved in the conversion of tryptophan to niacin. Also, riboflavin and niacin are involved in the conversion of folic acid to its coenzyme form. The conversion of vitamin B_{12} to its active form involves many nutrients, which include niacin, riboflavin, and manganese. Thus, many vitamins, including vitamin A and most of the B-vitamins, contribute to cellular antioxidant defense by being functionally involved in the repair or replacement of damaged molecules and cells.

REFERENCES

1. F. F. Ahmad, D. L. Cowan, and A. Y. Sun. Potentiation of ethanol-induced lipid peroxidation of biological membranes by vitamin C. *Life Sci.* **43**, 1169–1176 (1988).
2. S. D. Aust, L. A. Morehouse, and C. E. Thomas. Role of metals in oxygen radical reactions. *J. Free Rad. Biol. Med.* **1**, 3–26 (1985).

3. M. A. Babizhayev. The biophase effect of calcium on lipid peroxidation. *Arch. Biochem. Biophys.* **266,** 446–451 (1988).
4. A. Ballin, E. J. Brown, G. Koren, and A. Zipursky. Vitamin C-induced erythrocyte damage in premature infants. *J. Pediatr.* **113,** 114–120 (1988).
5. E. F. Barber and R. J. Cousins. Induction of ceruloplasmin synthesis by retinoic acid in rats: influence of dietary copper and vitamin A status. *J. Nutr.* **117,** 1615–1622 (1987).
6. L. R. C. Barclay. The cooperative role of glutathione with a lipid-soluble and a water-soluble antioxidant during peroxidation of liposomes initiated in the aqueous phase, and in the lipid phase. *J. Biol. Chem.* **263,** 16138–16142 (1988).
7. J. C. Bauernfeind. The safe use of vitamin A: a report of the International Vitamin A Consultative Group, The Nutrition Foundation, Washington, DC, 1980, p. 44.
8. A. Bendich and L. Langseth. Safety of vitamin A. *Am. J. Clin. Nutr.* **49,** 358–371 (1988).
9. R. E. Beyer, K. Nordenbrand, and L. Ernster. The function of coenzyme Q in free radical production, and as an antioxidant: A review. *Chemica Scripta* **27,** 145–153 (1987).
10. K. R. Bridge and K. E. Hoffman. The effects of ascorbic acid on the intracellular metabolism of iron and ferritin. *J. Biol. Chem.* **261,** 14273–14277 (1986).
11. G. W. Burton and K. U. Ingold. Autooxidation of biological molecules. 1. The antioxidant activity of vitamin E and related chain-breaking phenolic antioxidants in vitro. *J. Am. Chem. Soc.* **103,** 6472–6477 (1981).
12. G. W. Burton and K. U. Ingold. Beta-carotene: an unsual type of lipid antioxidant. *Science* **224,** 569–573 (1984).
13. G. L. Catignani, F. Chytil, and W. J. Darby. Vitamin E deficiency: immunochemical evidence for increased accumulation of liver xanthine oxidase. *Proc. Natl. Acad. Sci. USA* **71,** 1966–1968 (1974).
14. C. K. Chow. Vitamin E in blood. *World Rev. Nutr. Dietet.* **45,** 133–166 (1985).
15. C. K. Chow. Interrelationship of cellular antioxidant defense systems. In *Cellular Antioxidant Defense Mechanisms*, vol. 2, C. K. Chow, ed., CRC, Boca Raton, FL, 1988, pp. 217–237.
16. M. Comporti. Lipid peroxidation and cellular damage in toxic liver injury. *Lab. Invest.* **53,** 599–623 (1985).
17. H. Esterbauer, H. Zollner, and R. J. Schaur. Hydroxyalkenals: cytotoxic products of lipid peroxidation. *ISI Atlas Sci. Biochem.* **1,** 311–317 (1988).
18. W. E. Fahl, N. D. Lalwani, T. Wantanabe, S. K. Goel, and J. K. Reddy. DNA damage related to increased hydrogen peroxide generation by hypolipidemic drug-induced liver peroxisomes. *Proc. Nat. Acad. Sci. USA* **81,** 7827–7830 (1984).
19. J. C. Fantone and P. A. Ward. Role of oxygen-derived free radicals and metabolites in leukocyte-dependent inflammatory reactions. *Am. J. Pathol.* **107,** 395–418 (1982).
20. C. S. Foote. Photosentized oxidation and singlet oxygen: consequences in biological systems. In *Free Radicals in Biology*, W. A. Pryor, ed., vol. 2, Academic, New York, 1976, pp. 85–133.
21. I. Fridovich. Superoxide radical: an endogenous toxicant. *Ann. Rev. Pharmacol.* **23,** 239–257 (1983).
22. K. Gohil, L. Rothfuss, J. Lang, and L. Packer. Effect of exercise training on tissue vitamin E and ubiquinone content. *J. Appl. Physiol.* **63,** 1638–1641 (1987).

23. G. W. Grams, and K. Eskins. Dye-sensitized photooxidation of tocopherols. Correlation between singlet oxygen reactivity and vitamin E activity. *Biochemistry* **11**, 606–611 (1972).

24. J. M. C. Gutteride and B. Halliwell. The antioxidant proteins of extracellular fluids. In *Cellular Antioxidant Defense Mechanisms*, Vol. 2, C. K. Chow, ed., CRC, Boca Raton, FL, 1988, pp. 1–23.

25. B. Halliwell and J. M. C. Gutteridge. Oxygen toxicity, oxygen radicals, transition metals, and disease. *Biochem. J.* **219**, 1–14 (1984).

26. B. Halliwell and J. M. C. Gutteridge. Oxygen free radicals and iron in relation to biology and medicine: some problems and concepts. *Arch. Biochem. Biophys.* **246**, 501–514 (1986).

27. B. Halliwell and G. M. C. Gutteridge. Iron and free radical reactions: two aspects of antioxidant protection. *Trends Biochem. Sci.* **11**, 372–375 (1986).

28. B. Halliwell, M. Wasil, and M. Grootveld. Biologically significant scavenging of the myeloperoxidase-derived oxidant hypochlorous acid by ascorbic acid. *FEBS Lett.* **213**, 15–18 (1987).

29. E. W. Kellogg and I. Fridovich. Superoxide, hydrogen peroxide, and singlet oxygen in lipid peroxidation by a xanthine oxidase system. *J. Biol. Chem.* **250**, 8812–8817 (1975).

30. S. J. Klebanoff. Oxygen metabolism and toxic properties of phagocytes. *Ann. Intern. Med.* **93**, 480–489 (1980).

31. K. J. Kunert and A. L. Tappel. The effect of vitamin C on in vivo lipid peroxidation in guinea pigs as measured by pentane and ethane production. *Lipids* **18**, 271–274 (1983).

32. C. Little and P. O'Brien. An intracellular GSH-peroxidase with a lipid peroxide substrate. *Biochem. Biophys. Res. Comm.* **31**, 145–150 (1968).

33. J. A. Lucy. Functional and structural aspects of biological membranes: a suggested structure role of vitamin E in the control of membrane permeability and stability. *Ann. NY Acad. Sci.* **203**, 4–11 (1972).

34. J. F. Mead. Free radical mechanisms of lipid damage and consequences for cellular membranes. In *Free Radicals in Biology*, vol. 1, W. A. Pryor, ed., Academic, New York, 1976, pp. 51–68.

35. E. Niki. Antioxidants in relation to lipid peroxidation. *Chem. Phys. Lipids* **44**, 227–253 (1987).

36. H. Nohl and W. Jordan. The involvement of biological quinones in the formation of hydroxyl radicals via the Haber-Weiss reaction. *Bioorganic Chem.* **15**, 374–382 (1987).

37. J. A. Olsen. Vitamin A. In Nutrition Reviews' Present Knowledge of Nutrition, 5th Ed., The Nutrition Foundation, Washington, D.C., 1984, pp. 176–191.

38. D. T. Sawyer and J. S. Valentine. How super is superoxide? *Acc. Chem. Res.* **14**, 393–400 (1981).

39. D. Sklan. Vitamin A absorption and metabolism in the chick: response to high dietary intake to tocopherol. *Brit. J. Nutr.* **50**, 401–407 (1983).

40. D. Sklan. Vitamin A in human nutrition. *Prog. Food Nutr. Sci.* **11**, 39–55 (1987).

41. D. Sklan and S. Donoghue. Vitamin E response to high vitamin A in the chick. *J. Nutr.* **112**, 759–765 (1982).

42. M. K. Soliman. Vitamin-A-uberdosierung. I Mogliche teratogene wirkungen. *Int. Z. Vitam. Ernahrungsforsch (Beih)*, **42**, 389–393 (1972).

43. P. Starke and J. L. Farber. Ferric iron and superoxide ions are required for the killing of cultured hepatocytes by hydrogen peroxide. *J. Biol. Chem.* **260**, 10099–10104.

44. T. Udomratn, M. H. Steinberg, G. D. Campbell Jr., and F. J. Oelshegel Jr. Effects of ascorbic acid on glucose-6-phosphate dehydrogenase-deficient erythrocytes: studies in an animal model. *Blood* **49,** 471–475 (1977).

45. F. Ursini, M. Maiorino, and C. Gregolin. The selenoenzyme phospholipid hydroperoxide glutathione peroxidase. *Biochim. Biophys. Acta* **839,** 62–70 (1985).

46. W. C. Willet, M. J. Stampfer, B. A. Underwood, J. O. Taylor, and C. H. Hennekens. Vitamins A, E, and carotene: effects of supplementation on their plasma levels. *Am. J. Clin. Nutr.* **38,** 559–566 (1983).

47. G. Wolf. Multiple function of vitamin A. *Physiol. Rev.* **64,** 873–937.

From: *Trace Elements, Micronutrients, and Free Radicals* • Ed.: I. E. Dreosti • ©1991 The Humana Press Inc.

CHAPTER 7

Free Radical Pathology and the Genome

Ivor E. Dreosti

ABSTRACT

Because of the ubiquity of oxygen in the metabolism of most forms of life, oxygen-derived free radicals occur widely in biological tissue. Cellular damage arises only when prooxidant free radical flux exceeds the natural antioxidative defense mechanisms, or when these systems are hypoeffective. Genomic damage appears to derive principally from attack by the highly reactive hydroxyl radical, against which several micronutrients, including vitamins C and E and zinc, could be expected to be protective. Since generation of the hydroxyl radical must occur intranuclearly if genome damage is to be effected, iron-catalyzed production of this species from hydrogen peroxide becomes of central interest, as also does the capacity of the putative nuclear antioxidants to prevent it. Zinc attracts particular attention in this regard because of its postulated ability to displace redox-active iron from chromatin on the one hand, and to induce the hydroxyl radical scavenger metallothionein on the other.

1. INTRODUCTION

All major classes of biological macromolecules are vulnerable to free radical damage, and genomic DNA is no exception. The consequences to the cell of chromosome injury are variable and are largely dose-related, although significant damage can occur even at very low clastogenic

149

exposures, following a single random genetic modification referred to as a stochastic event. Generally however, low levels of genome damage are effectively repaired and a viable, nonmutated cell survives. Massive genome damage overwhelms the capacity of the cell for DNA repair and is cytotoxic, thereby eliminating the potential for mutagenesis. Intermediate degrees of chromosomal injury elicit DNA repair processes of increasingly less fidelity, with an attendant dose-related likelihood of mutagenesis and the parallel event of carcinogenesis. Substances that protect against genome damage are, therefore, generally antimutagenic, as also paradoxically are many compounds that diminish the rate of DNA repair and/or cell division.

Conditions leading to chromosomal injury include a number of chemical and physical factors that affect the genome in a variety of ways. Lately, particular interest has focused on the part played by oxygen-derived free radicals in relation to DNA damage, and on the capacity exhibited by some metals to enhance the levels of reactive oxygen species, whereas several other trace elements and vitamins tend to reduce them. The present paper examines the involvement of free radicals in DNA damage against the background of iron as a prooxidant, and the cellular antioxidant systems comprised essentially of vitamin and mineral micronutrients. Genetic damage, DNA repair, mutagenesis, and carcinogenesis are discussed in relation to free radical flux in the nucleus, and to the protection afforded by the antioxidant micronutrients relevant to that cellular compartment.

2. FREE RADICALS OF BIOLOGICAL SIGNIFICANCE TO THE GENOME

2.1. Free Radical Species

Free radicals may be defined as any chemical species that has one or more unpaired electrons (33). They occur widely in biological tissue, often as endogenous intermediates of normal physiological processes, although sometimes, they are generated endogenously in response to an exogenous stimulus, or they may be introduced into the tissue entirely from an exogenous source. Many metabolic processes that proceed via free radical intermediates contribute to endogenous sources and include the mitochondrial electron transport chain, the cytochrome-p 450 system, certain oxygenases, and some dehydrogenases, cyclooxygenases, lipoxygenases, and peroxidases, as well as the phagocytic destruction of bacteria by neutrophils and the autooxidation of a wide variety of soluble cell components (e.g., hydroquinones, catecholamines, flavins, and chelated Fe^{2+} iron) (28,30). Some endogenous free radical production derives from the metabolism of xenobiotics (e.g., pesticides, solvents, anesthetics, aromatic hydrocarbons, antineoplastic agents, and antibiotics), usually through the cytochrome-p 450 and related systems (10,28). Other exogenous factors responsible for the production of endogenous free radicals include ultraviolet, electromagnetic (X and γ-rays), and particulate (electrons, protons, neutrons, and so on) irradiation, and hyperbaric oxygen (10,66). Electromagnetic radiation produces free radicals either by transferring energy directly to biological molecules, thereby making them free radicals, or indirectly by the action of primary radicals (e_{aq}, OH· and H·) derived from the radiolysis of water acting in turn on other cellular constituents to promote free radical mediated macromolecular degradation (66). Exogenous radicals introduced into the cell include small carbon and oxygen-centered species, that occur, for example, in smog and cigarette smoke (7).

Because aerobic metabolism predominates in most forms of life, and because oxygen readily accepts one-electron transfers, oxygen-centered free radicals are the most common mediators of cellular free radical reactions (28), although hydrogen, carbon, chlorine, sulphur, and nitrogen-centered radicals do occur in biological tissue, and are of toxicological relevance (34). The principal oxygen-centered radicals include, but are not limited to, the inorganic superoxide (O_2·), hydroxyl (OH·), and hypochlorite (Cl O·) species, as well as the organic alkoxy (RO·) and peroxy (ROO·) radicals (34,43). Reactive molecules, such as hydrogen peroxide (H_2O_2) and singlet oxygen (O_2), are not free radicals but, nevertheless, contribute to cellular peroxidation and to free radical related damage (12,43). In fact, the oxygen molecule itself has two unpaired electrons and, by broad definition, is a free radical although it does not behave as one, owing to restrictions imposed by the similarity of the spin quantum number of these two electrons (35,66).

2.2. Transition Metals and Free Radicals

Transition metals, for example, copper and iron, in their lower oxidation states participate readily in the univalent reduction of oxygen, and thereby generate a number of oxygen-centered free radicals, especially highly reactive OH·, that will attack almost any biological molecule with which it comes into contact, causing strand breaks in DNA and hydroxylation of purine and pyrimidine bases, as well as attacking adjacent and nearby membrane lipids and cellular proteins (33,35). Although cuprous ions (33–35,63,75,76), and possibly manganese (5), do feature in these reactions, and indeed have a high rate constant for the formation of OH· from H_2O_2 in vitro, divalent iron is much more important pathophysiologically, as copper-catalyzed oxidations tend to be slower than those of iron (75), and copper ions bind widely to cellular proteins, especially albumin, which in turn behave as sacrificial antioxidants being oxidized themselves by OH· before other cellular damage can occur (35). Manganese may be of some importance because of its capacity to form relatively stable reactive compounds with oxygen, which in turn can transfer the oxidative damage to more distant sites (5).

It is now apparent that within the cell, complex mechanisms exist for keeping free divalent cations, especially iron, bound in elaborate macro-molecular structures, and for protecting vital sites sensitive to iron-catalyzed oxidations from those ions by sequestration within complicated macromolecules, or by complexation with zinc, which is catalytically inactive. (75,76). Nevertheless, cells do contain a small pool of low-mol-mass iron that can be added to by oxidant stress and tissue damage, and that probably represents the main catalyst for oxidative damage, al-though several forms of bound iron are also capable of facilitating the production of OH· from H_2O_2, and lipid alkoxy (RO·) and peroxy (ROO·) radicals from lipid peroxides (33,34).

Although in the past, most attention concerning transition metals and cellular peroxidative damage has tended to focus on membrane lipids, increasing evidence now points to a role for iron, and possibly copper, in free radical related genome damage. Thus, studies in vitro have shown that DNA that has been purified free of metals barely reacts with H_2O_2, or with O_2· generated by pulse radiolysis, whereas the little O_2·-related damage that does occur can be prevented with superoxide dismutase (8). Accordingly, it has been suggested that chromosomal damage in vivo may be related, at least in part, to the generation of OH· at specific sites on the DNA molecule itself, that arise from catalysis by transition metals bound directly to DNA or to chromatin protein, or possibly, liberated within the nucleus by oxidative stress (27,28,75). The notion is supported by several studies pointing to the binding of ferrous iron to DNA (27,46,75,76), and to increased DNA damage (17,27,46) and cell killing (25,58) by H_2O_2 in cells exposed to supplementary iron, which is reduced if the iron is removed from the system by strong chelating compounds (25,46). Furthermore, protection against DNA damage is generally obtained by the removal of O_2· by superoxide dismutase and of H_2O_2 by catalase, or by the scavenging of OH· by superoxide dismutase.

The involvement of iron in DNA breakage is particularly well exem-plified in the action of the antitumor, antibiotics, adriamycin (49) and bleomycin (31,32) that bind to both DNA and to ferrous ions and, under aerobic conditions or in the presence of hydrogen peroxide, liberate OH· that provoke DNA strand scission. Similar iron-related oxidative DNA cleavage is evident in the experimental use of an EDTA/ferrous complex attached to a sequence-specific DNA-binding molecule, in order to create a predictably precise DNA affinity cleaving molecule (22).

2.3. Transition Metals and Genome Damage Without Free Radicals

Brief mention should be made here to the widely documented capac-ity of metals, especially those in the divalent and transition series, to bind to DNA and to act as mutagens, without necessarily involving free radical mechanisms (*15*). Thus, arsenic, beryllium, chromium, iron, copper, manganese, molybdenum, platinum, and selenium are mutagenic in prokaryotes, whereas beryllium, copper, iron, manganese, platinum, arsenic, cadmium, lead, nickel, chromium, silver, and zinc are mutagenic or carcinogenic in mammalian cells in culture (*15*,*63*). The mechanisms whereby these metals act as mutagens or carcinogens are not clear, but they all do exhibit an ability to induce one or more lesions in DNA, such as strand breaks and crosslinks, sister chromatid exchanges, and DNA repair synthesis (*15*). It is possible however to draw an interesting distinction between the toxic metal ions and normal essential trace metals. Chemical reactivity, as measured by the physiocochemical parameter of "softness," has been shown to correlate well with the LD_{50} of metal ions. The most toxic metal ions, because of their easily deformable outer electron shells, tend to form bonds of a more covalent nature, whereas essential ions are "hard" ions that form bonds that are essentially ionic, and can therefore be displaced from normal metal-mediated functions by the more stable structures formed with toxic metals (*15*).

Overall, in vitro genotoxic testing of about 25 different metals, or metal compounds of varying levels of carcinogenicity, suggest effects on the genome, which may include covalent metal binding to DNA, crosslink formation within DNA and protein, and impaired fidelity of DNA replication, transcription, and DNA repair (*16*,*53*), none of which involve oxygen free radicals, although this latter aspect of metal-related genomic damage is attracting increasing attention and is the subject of this chapter.

2.4. Ionizing Radiation and Free Radicals

Considerable similarity exists between the DNA damage associated with the biochemical production of OH· in the cell and that derived from exposure of biological tissue to ionizing radiations (X-rays, γ-rays, high energy electrons, and so on) that may disrupt DNA by direct action or indirectly through the formation of a variety of daughter radicals. Thus, DNA may be the primary target, with the result that it, or its constituent bases, become free radicals, or it may be affected secondarily through the radiolysis of water, which produces a variety of reactive species that subsequently attack and disrupt neighboring molecules, including DNA (*42*,*64*).

In Summary (*64*):

(1) Direct action: biomolecule $\xrightarrow{\text{irradiation}}$ biomolecule·

(2) Indirect action: $H_2O \xrightarrow{\text{irradiation}} OH^\cdot, e^-_{aq}, H^\cdot$

$OH^\cdot \xrightarrow{\text{biomolecule}}$ biomolecule·—OH

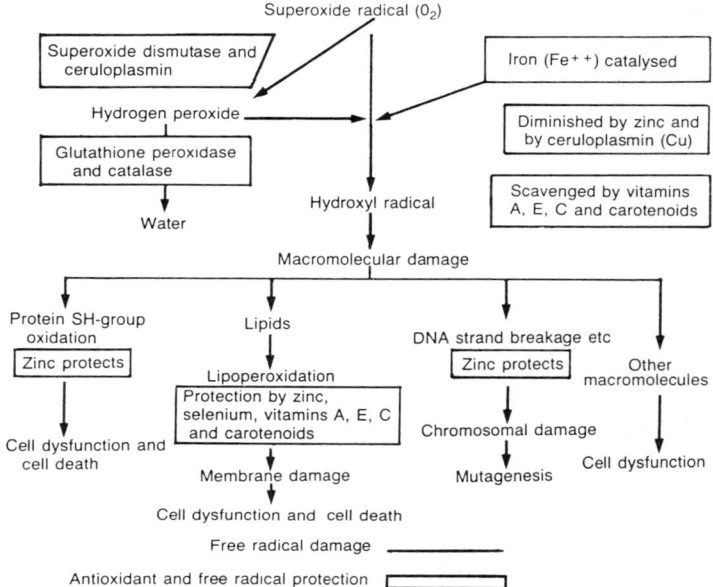

Fig. 1.   Cellular free radical damage and antioxidant defense mechanisms. GSH, reduced glutathione; SOD, superoxide dismutase; (□), antioxidants and free radical scavengers. (Reproduced with permission: I. E. Dreosti. Antioxidants, micronutrients versus free radicals. *Australian Family Physician* **17,** 684–686 (1988).)

2.5. Sites of Production of Free Radicals

Free radicals are produced throughout the cellular milieu, and occur in mitochondria, lysosomes, peroxisomes, the nuclear, endoplasmic reticular and plasma membranes, and within the cytosol.

Free radicals produced intramitochondrially would tend to be neutralized by internal antioxidant defenses, mainly in the form of superoxide dismutase (28). Free radicals produced elsewhere in the cell would encounter various components of the multilevel antioxidant and free radical scavenging defense system, thus, ensuring some measure of protection of cellular constituents against rampant free radical damage (Fig. 1).

However, in the case of free radicals derived within the nucleus (e.g., X-irradiation) or entering the nucleus from the cytoplasm or nuclear membrane (e.g., H_2O_2, O_2), DNA would be particularly vulnerable to oxidative damage. In this regard, both the endoplasmic reticulum and the nuclear membrane contain the cytochromes-p 450 and -b 5, that oxidize fatty acids, xenobiotics, and other substances, using NADPH or NADH as cofactors, often with a degree of uncoupling that diverts electron flow to dioxygen, and leads to the formation of O_2^- and H_2O_2 (28).

The question of which species of oxygen radical is most reactive and damaging within the cell seems generally to implicate OH, although this radical is so reactive (Half-life $= 10^{-9}$ s) that it will react with a bio-molecule before diffusing more than 2 or 3 molecular diameters from the site at which it is produced. The alkoxy radical (RO) is only slightly less reactive (Half-life 10^{-3} s), whereas the peroxy radical (ROO) with a lifetime of several seconds can diffuse considerable distances within the cell (57,74). Superoxide (O_2^-) and H_2O_2 are less reactive than OH and can migrate some distance away from where they are produced, and are therefore able to reach different parts of the cell where they may generate OH in a reaction catalyzed by free transition metal ions capable of undergoing redox changes (66,75). Hydrogen peroxide (H_2O_2) crosses cell membrane easily and could, for example, migrate into the nucleus where it might react with free redox metal ions in the nucleoplasm, or with Fe^{2+}-protein/DNA or other metal–macromolecule complexes to form OH (27,46,58). It has been suggested that Fe^{2+} bound to DNA is more stable than ionic Fe^{2+}, which rapidly (20 s) becomes oxidized to Fe^{3+} (27). In this way, some Fe^{2+}-complexes may enhance and prolong the effectiveness of iron in catalyzing the production of OH from H_2O_2, and consequently, may be more damaging to the cell than free ionic iron alone. The capacity of H_2O_2 to enter the nucleus and to produce OH, thereby inducing single strand breaks on chromosomal DNA makes H_2O_2 a good chemical mimetic of ionizing radiation (46). Superoxide does not easily cross cell membranes unless it can pass through a membrane channel (46), and in so doing, avoid reacting with constituent membrane molecules.

A major consideration in evaluating the potential for free radical damage at a particular point within the cell must, therefore, depend on the nearest site of production of the damaging species, as well as the likelihood of the free radicals so produced reaching a subcellular target before their reactive nature causes them to combine with other macro-molecules *en route*. Superoxide and H_2O_2, which can both diffuse considerable distances in the cell before undergoing conversion to more reactive OH, would appear to be principal players in this role, in the series of reactions that are summarized below:

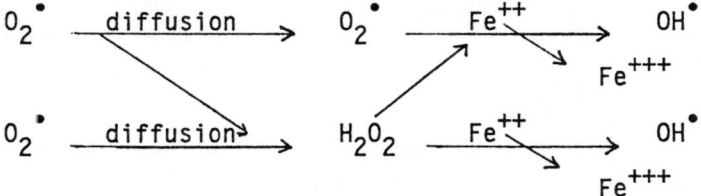

Thus, the probable, critical determinant of the extent of damage done by H_2O_2 and O_2^- is not so much the level at which they are

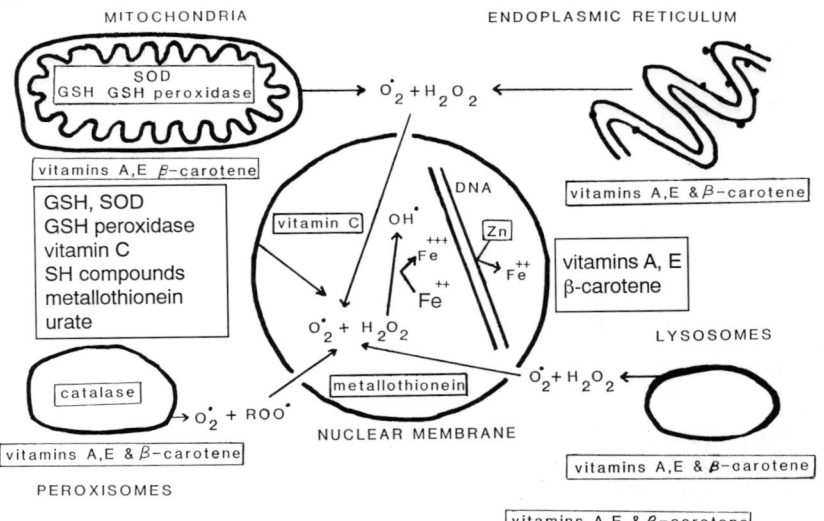

Fig. 2.   Subcellular location of free radical production and free radical defense systems in relation to the nucleus and chromosomal DNA.

produced but where, within the cell, the metal ion complexes occur that will catalyze the formation of OH·. If they are on DNA, then DNA strand scission may result, but if the metals are bound predominantly to bio-membranes, then they will give rise to lipid peroxidation and the products of lipoperoxidation (33). Some free radical damage "at a distance" may, in fact, derive from diffusion of the lipid peroxidation intermediates, notably RO· and ROO·, which react further to produce H_2O_2 and toxic aldehydes that also travel some distance from their source (28). Certainly, products of lipoperoxidation, such as malondialdehyde, have been reported to react with nucleic acids and to be mutagenic and carcinogenic (2,33,43,61), and site-specific cleavage of double-stranded DNA has been noted in relation to the hydroperoxide of linoleic acid (39).

2.6. Free Radicals and Genome Damage

Recognizing that OH· is probably the active oxygen species most responsible for free radical damage to the genome, and that OH; damage in biological tissue tends to be associated with the diffusion of $O_2·$, H_2O_2, and lipid peroxides to areas of high concentration of the redox-cycling divalent cations iron and copper, a picture of free radical attack on the genome emerges. The scenario involves diffusion into the nucleus of $O_2·$ and H_2O_2 produced in the cytosol, along with more of the same radical species and lipid peroxides arising from the nuclear membrane, until they come into contact with chromatin-bound metal ions that catalyze the production of OH· and initiate the attendant DNA damage (Fig. 2).

3. FREE RADICALS AND DNA DAMAGE

3.1. Type of DNA Damage

DNA can be damaged in many ways, ranging from chemically in-duced alkylation of DNA bases, through light-mediated intrastrand crosslinks (e.g., pyrimidine dimers) and interstrand crosslinks (e.g., psoralen-type bridges) to the wide variety of anomalies caused by the action of free radicals. This latter category includes base alteration, base detachment from an intact phosphodiester backbone to yield an apurinic site, intercalation, intra and interstand crosslinks, DNA-protein cross-links, intraprotein crosslinks, and single and double stand breaks that, if present during DNA replication, may result in nontemplate incorpora-tion of nucleotides and spontaneous or induced mutagenesis (64,65).

Damage arising from ionizing radiation tends to be unevenly distrib-uted, and appears in clusters around the ionizing events, which are known as spurs. Chemically-generated damage is more uniform in distri-bution, except where the chemical generator of free radicals is attached to DNA (e.g., redox-cycling metals), when the anomalies follow the pattern of spurs (64).

Reaction of DNA with free radicals occurs at five points: the four bases, thymine, cytosine, adenine, and guanine, and the sugar moiety, deoxyribose. Phosphate is relatively unreactive towards free radicals and is not significantly involved in DNA damage (20). Oxygen derived free radicals can either oxidize or reduce nucleotide bases, but deoxyribose only undergoes hydrogen ion abstraction, resulting in a deoxyribose radical in DNA, which possibly leads to the loss of a base, a strand break, or internal cyclization with a base attached to it (20). In a broad sense, all these small radicals confer upon the entire DNA molecule the character of a radical, but in practice, most attention is paid to the radical nature of the small component compound (64). Formation of pyrimidine (mainly thymine) and purine (mainly guanine) radical species accompany radia-tion of DNA, and result in electron trapping at thymine $(T.^-)$ and electron loss at guanine $(G.^+)$ (65). Details of these free radical mecha-nisms, the accompanying base damage, and the free radical forms and transient species have been extensively described in a treatise elsewhere (45). Electron spin resonance evidence indicates that both $G.^+$ and $T.^-$ primary centers result in permanent damage to DNA, with a relatively high incidence of strand breaks arising from what probably involves hydrogen abstraction from deoxyribose, position C-2' or C-4', and het-erolytic splitting of the phosphoric acid ester bond at the C-3' or C-5' position (52). In the presence of oxygen, many carbon-centered radicals are converted into their corresponding peroxyl radicals, especially py-rimidines and the sugar moiety. Thus, primary base peroxyl radicals appear to play a central role in attacking the deoxyribose backbone to form sugar radicals that take up oxygen leading to strand breakage and base release (70).

The experimental observation that electrons generated within histone proteins by radiation are effectively transmitted to DNA, leading to an enhanced initial yield of G.$^+$ and increased sensitization of radiation damage (20), is particularly important because of the apparent association of metal ions with chromatin proteins. Thus, it seems that direct damage to DNA, as well as that proceeding via neighboring proteins, may rank equally in importance with the indirect mechanisms of radiation damage.

3.2. DNA Damage During Transcription and Protection by Proteins

Many defense mechanisms exist to protect the cell against reactive oxygen species, which include the several primary antioxidants and secondary free radical scavengers discussed elsewhere in this volume (Fig. 1). For DNA however, further levels of defense apply, in that it is compartmentalized away from the organelles (mitochondria, peroxisomes, endoplasmic reticulum) where most radicals are generated, and it is surrounded by a protective layer of histones—at least, when it is not replicating. Furthermore, most types of damage to DNA can be effectively repaired by DNA repair mechanisms (3). It is of interest that whereas actively transcribing regions of DNA are less protected by chromatin proteins and, therefore, more vulnerable to radiation damage, digestion of chromatin by nuclease enzymes yields soluble nucleoproteins containing elevated levels of zinc and copper (60), which suggests that these metal–ion-enriched areas appear to be associated with those parts of the chromosome actively transcribing RNA and at risk from free radical damage.

3.3. Assessment of DNA Damage

Free radical damage to DNA may be detected by chemical, physical, enzymatic, and immunochemical methods. Chemical methods are generally relatively insensitive, but can be made to be highly specific (3). A particularly popular test for genotoxicity involves measurement of inhibition of semiconservative synthesis of DNA by cellular uptake and incorporation of radioactive thymidine (15). Unfortunately, the method is subject to considerable variability relating to nucleotide pool size, and may be influenced by so many cellular disturbances that it provides limited information concerning the actual mechanism of genome damage. More recently, a specific chemical assay has been developed for three free radical damage products of DNA, namely, thymine glycol, 5-hydroxymethyluracil, and 8-hydroxyguanine, all of which can be measured in urine at background levels of damage from normal aerobic metabolism (3).

Physical assays, that detect actual DNA damage are currently widely used, and include the recently developed technique for determining single strand breaks of 2–5×10^8 dalton in DNA by centrifugation of DNA in alkaline sucrose gradients (15). More recently, an improved technique involving alkaline elution can measure DNA strands 5×10^8–10^{10} dalton in size and, when used in conjunction with proteases or X-irradiation, can establish the extent of DNA–protein and DNA–DNA crosslinks (15). Both alkaline sucrose gradient centrifugation and alkaline elution estimate the amount of single and/or double strand breaks induced in DNA, but the alkaline elution method is more sensitive to lower levels of DNA damage (1 single strand break/10^9 dalton vs 50 single strand breaks/10^{10} dalton), and is the preferred technique (15). Neutral conditions can be used to distinguish double strand breaks, but since most agents primarily cause single strand scission (double strand breakage occurring mainly during repair), alkaline conditions are more widely used (15). Generally however, these techniques are mainly suitable for detecting low levels of DNA damage in small prokaryote genomes, but are of limited usefulness with mammalian cells (3).

In most cases, a potential consequence of DNA damage may involve changes of chromosome structure and attendant chromosome aberrations, which can be assessed by conventional chromosome spreads, or by the quicker and less tedious technique of sister chromatid exchanges that arise during DNA repair. However, no technique is faultless and in some cases, severe chromosomal anomalies are not accompanied by any significant sister chromatid exchanges. Most recently, the newly proposed and rapid micronucleus test scores chromosomal aberrations in relation to nuclear fragments arising from damaged chromosomes that are excluded from daughter nuclei following division of a DNA damaged nucleus (60).

DNA damage of particular relevance to the initiation of carcinogenesis often seems to involve covalent reaction of the carcinogen with DNA to produce DNA adducts that, in themselves, provide a useful measure of DNA damage, as they can be quantified with radioactive carcinogens or by a variety of fluorescent and immunological assays (60).

4. DNA REPAIR, MUTAGENESIS, AND CARCINOGENESIS

The functional significance of DNA damage depends on its severity, whether in the first instance it is sufficiently serious to cause cell death or, if it is less severe, whether it may lead to mutagenesis, teratogenesis, or carcinogenesis. Overall, most DNA damage associated with genotoxic events is repaired in the cell to some extent, although the operation is less effective when chromosome damage is extensive, as more error-prone repair systems may be involved and DNA replication is sometimes initiated before adequate repair has been effected (21,64). Lack of repair itself may be highly carcinogenic, especially in certain regions of chromosomal DNA (6). Several trace elements, notably mercury and cadmium, interfere with repair processes, and may thus contribute to greater cell death following severe genomic damage, that in turn may, under certain circumstances, actually diminish the carcinogenic potential of the treatment owing to reduced cell survival (6,64). DNA damage is not in itself synonymous with mutagenesis, in the extreme, it is cytotoxic, but at intermediate levels, the potential exists for incorrect repair and accompanying genetic modification (52).

Three possible mechanisms have been proposed whereby free radicals may lead to mutagenesis and carcinogenesis (57). Suggested mechanisms involve: (1) activation of a carcinogen by a free radical dependent process to an electrophylic epoxide that reacts directly with DNA (33), (2) radical activation of a carcinogen to a diol or quinone that can bind to DNA (30), and (3) the conversion of carcinogens to diols or quinones that do not bind to DNA but instead redox cycle to produce oxy radicals and H_2O_2 that damage DNA (28). It is the latter mechanism that is currently receiving particular attention as it appears to operate on a very much broader biochemical front, involving other redox agents (e.g., Fe^{2+}, Fe^{3+}) and other generators of oxy radicals.

Recently, the likelihood has been recognized that damage to DNA may be critical, both in the initiation and promotion stages of carcinogenesis (10,40,43). Active oxygen species, because of their ubiquity in aerobic tissue, have become of central research interest, since their involvement in damage to other biological macromolecules (e.g., membrane lipids, enzyme proteins) suggests that they may be implicated in carcinogenesis at levels other than mutagenesis, that possibly encompass disturbances in cellular control mechanisms, especially relevant to the promotion and progression of tumor growth (11,78).

Evidence of a role for active oxygen in cancerogenesis derives in part from the observation that several human hereditary diseases that are distinguished by an increased incidence of spontaneous chromosomal breaks and cancer, are characterized also by abnormal intracellular oxygen metabolism (10). In addition, many prooxidant conditions (e.g., hyperbaric oxygen, peroxides, xenobiotic metabolism of bleomycin, daunorubicin, polycyclic hydrocarbons, and so on, aerobic ionizing radiation) are carcinogenic, whereas many antioxidants (e.g., butylated hydroxytoluene, butylated hydroxyanisole, vitamins A, C, and E, β-carotene, superoxide dismutase, catalase, glutathione peroxidase, urate, glutathione, and other sulphydryl-containing compounds) have been reported to act as anticarcinogens (10).

5. PROTECTION OF THE GENOME BY ANTIOXIDANTS

Protection of the genome against free radical damage will presumably involve similar cellular defense systems to those that operate elsewhere in the cell, except that there may be substantial differences in the relative activity of certain systems between subcellular compartments and between tissues (Fig. 2).

In general, damage to DNA appears to arise primarily from $OH^{\cdot}$. However, this reactive species is so short-lived that, with the exception of ionizing radiation or other systems capable of generating $OH^{\cdot}$ directly within the nucleus, most nuclear $OH^{\cdot}$ must arise indirectly from less potent species that can diffuse into the nucleus, and that probably include mainly $O_2^{\cdot}$ and H_2O_2 generated in the cytoplasm and on the endoplasmic reticular and nuclear membranes by various oxygenases and the cytochrome-p 450 system (*17*). Thus, cytoplasmic catalase and glutathione perioxidase, as well as lipophylic free radical scavengers in the nuclear membrane (e.g., vitamins A, E, and β-carotene), water soluble vitamin C in the nucleus, and chromatin-bound zinc may be expected to confer protection to nuclear DNA when replete or in superabundance, and to enhance susceptibility to free radical damage when deficient.

5.1. Vitamins C and E

In the context of genome protection, dietary variations of both vitamins C and E are known to affect the LD50 in experimental animals exposed to X-irradiation (*50,55*) and appear, therefore, to play an important role in defense against radiation-induced free radical damage. Since the environment within the nucleus is essentially aqueous and possesses few antioxidant enzyme systems, vitamin C, which acts as the principal water-soluble antioxidant in many other fluid compartments (*43*), and which is present also in the nucleus (*48*), could be expected to operate to some extent in this environment as a scavenger of superoxide and hydroxyl radicals (*28,36*). Vitamin E, which is probably the principal lipophylic defense compound in the cell, is also located in the lipid bilayer of the nuclear membrane, where it would scavenge superoxide and hydroxyl radicals produced by membrane-bound enzyme systems, or derived from reactions in the aqueous phase (*28*). The established responsiveness of membrane vitamin E levels to the dietary intake (*43*) suggests that the effectiveness of this putative nuclear antioxidant may also be amenable to manipulation by dietary means.

5.2. Zinc

The trace element, zinc, deserves consideration as it too has been found to effectively reduce the lethality of X-irradiation in rats (*26,44*). An important and possibly pivotal role has recently emerged for zinc in displacing bound iron from radical-sensitive macromolecules, thereby reducing the iron-catalyzed production of highly reactive $OH^{\cdot}$ from less active $O_2^{\cdot}$ (*29,75*). DNA purified to be free of iron reacts very slowly with superoxide, which suggests that DNA damage in vivo may be as a result of a site-specific generation of OH' on the DNA itself, by closely associated redox-cycling metal ions (34). Substantial evidence now indicates zinc to be important in stabilizing DNA (57,58), and indeed, it appears to be retained longer in the nucleus than in any other cell compartment following administration of radiolabeled ^{65}Zn (72). The possibility, therefore, exists that as intracellular levels of zinc increase, more iron will be displaced from binding to chromosomal nucleoproteins, and less OH'-driven base damage and strand breakage of DNA will occur. The notion is further supported by the greater binding affinity of zinc, compared with iron in relation to most cellular proteins, especially those rich in thiol groups, and with tetrahedral binding sites (73,75).

The reported capacity of zinc to inhibit microsomal NADPH oxidation and NADPH-dependent cytochrome-*c* reductase activity (18) may also relate directly to reduced reduction of OH' in the endoplasmic reticulum and the nuclear membrane, and is a consideration that should not be overlooked with respect to the antioxidant activity of zinc in the cell.

5.3. Metallothionein

Metallothioneins (mt) are a class of proteins (mol wt approx 6500 dalton) consisting of a single polypeptide chain of 61–62 amino acids containing 20 cystein residues, but no aromatic amino acids or histidine, and which contain seven bivalent metal ions (Zn,Cu,Cd,Hg) bound through metal–thiolate linkages (37). The physiological function of metallothionein is not clear, and although it appears to be principally involved in detoxification of heavy metals and the metabolism of several essential trace elements, its induction by other stimuli, including X- and UV-irradiation and several anticarcinogenic therapeutic agents, suggests it may also function metabolically to scavenge OH' and $O_2^{.-}$ (4,14, 18,38,69,71). Scavenging of OH' by mt appears to be highly effective, and involves primarily the cysteinyl thiolate group (71), that is 20 times more abundant in mt than in glutathione, which itself is recognized to be a major cellular defense system against oxygen radicals (9). Such scavenging appears to involve loss of metal ions from mt, and a concomitant formation of disulphide bridges that can be restored by treatment of the mt with reduced glutathione and zinc ions, suggesting that in cellular systems at least, radiation damaged mt may be repaired by a glutathione-dependent mechanism (69).

Recently, zinc metallothionein (Zn-mt) has been reported to be induced in cells in culture (14), and in the livers of rats (69) subjected to X-irradiation. Furthermore, it appears that the induction of mt in cells in culture by the administration of cadmium (4), or in rats in vivo by zinc (69) confers on the organism some measure of protection against cell lethality associated with X-irradiation. Also, induction of mt by zinc in cultured hamster cells suppressed the cytotoxicity of *t*-butyl hydroperoxide, adding further support to the notion that mt acts as a scavenger for the reactive radical species that are formed from oxygen in an iron-mediated manner (*54*). Thus, it appears that mt may react sufficiently rapidly with OH˙ as a sacrificial target to reduce cytotoxic oxidative damage elsewhere in the cell (*69*). In addition or alternatively, mt may act as a hydrogen donor in the repair of DNA and may confer protection against OH˙ more through an enhancement of rapid repair mechanisms than by diminishing DNA damage (*1*).

Central to speculation concerning the role of mt as a scavenger of potentially clastogenic OH˙ must be the consideration whether the protein is located to any extent within the nucleus, where chromosome damage associated with short-lived OH˙ would have to occur. Several reports point to the accumulation of mt-inducing metals on the acidic nonhistone chromatin proteins leading to induced synthesis of the protein (*13,59*). Furthermore, whereas generally a cytosolic protein, mt induced by the divalent metals, zinc, copper, cadmium, and mercury, has been demonstrated immunohistochemically in humans and rats to be present also to some extent in the nuclei of livers and kidneys (*13,51*). Recently, contrary to general belief, iron has been shown to induce hepatic mt in chick hepatocytes, despite a low affinity of iron for the protein. However, induction required the iron to be injected intraperitoneally, which could have led to the synthesis of mt subsequent to physiological stress, as occurs with heat, cold, food restriction, and infection (*56*).

Overall, significant evidence has accumulated that mt plays an important role in the protection of cells against free radicals released either from ionizing radiation (*4,69*) or during cellular oxidative metabolism (*38,54*). Whereas a role for mt as a sacrificial target for oxidative damage appears likely, its contribution to maintaining the sulphydryl status of the cell also warrants consideration (*69*). Furthermore, the possibility cannot be excluded that the protection afforded by mt may arise from enhanced chemical repair by ablation of DNA radical sites that would otherwise be sensitive to oxygen-related strand scission, as has been observed with other thiol-containing compounds (*1,47,67*) and other hydrogen or electron donors (*15,20*).

6. MICRONUTRIENTS AND MICRONUCLEI

A common consequence of free radical related damage to DNA involves alterations in chromosome structure and accompanying aberrations in chromosome patterns. The appearance of micronuclei (Fig. 3) in damaged cells has recently attracted increasing interest as these cytoplasmic bodies have been demonstrated to occur in a largely dose-related manner following exposure to X-irradiation and to several chemical free radical generating systems (*15*).

Micronuclei are extranuclear DNA-containing bodies that result from chromosome or chromatid fragments, as well as aberrant chromo-

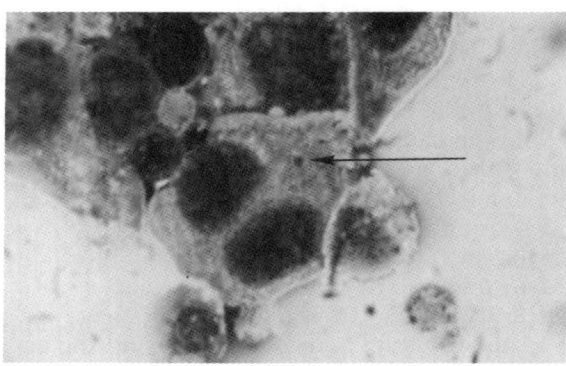

Fig. 3. Micronucleus (arrowed) in murine splenocytes obtained by cyto-kinesis block after X-irradiation with 1.5 Gray.

somes that have been excluded from the daughter nuclei following cell division. They reflect, in particular, DNA damage in which chromosome fragments have become detached from the kinetochore, and are, therefore, good indicators of the extent of double stranded DNA scission. The technique provides a useful quantitative assessment of genomic damage and the accompanying potential risk of cancer (19,68). The recently developed cytokinesis-block, micronucleus technique greatly enhances the statistical precision and sensitivity of the method, and provides a meaningful and rapid assessment of genotoxicity (24).

Studies recently performed in our laboratory using this technique with mice injected daily with 300 µg of zinc for two weeks before X-irradiation with 1.5 Gray, indicated a significant ($p < 0.05$) reduction of about 27% in the number of micronuclei counted in cultured splenocytes after treatment with the mitogen concanavalin A and the cytokinesis blocker cytochalasin B (23). This use of the micronucleus technique in an experimental nutritional setting has provided data that further support the view that zinc may play a significant role in the cellular defense mechanisms against free radical attack, and highlight, in particular, the apparent role of zinc as a nuclear antioxidant.

7. CONCLUSIONS

Extensive macromolecular damage associated with uncontrolled free radical activity in the cell results in cellular dysfunction at all levels, including chromosomal aberrations and genome modification. Antioxidant mechanisms within the cell provide a multilevel defense system against prooxidant free radical attack, and generally incorporate micronutrients as key components. The distribution of these compounds in the cell is not uniform, nor is the efficacy with which they act as antioxidants.

Indeed, with respect to the nucleus and its complement of chromatin, some antioxidants could be argued, at least on theoretical grounds, to provide better protection than others.

Zinc, because of its potential to displace redox-active iron from nuclear macromolecules, and because of its ability to inhibit the activity of cytochrome-p 450 and to induce metallothionein, appears to be especially well positioned to act pivotally in the protection of the cell against OH⁻-initiated genome damage.

REFERENCES

1. G. E. Adams. Radiation chemical mechanisms in radiation biology. *Adv. Radiat. Chem.* **3**, 125–129 (1972).
2. B. N. Ames. Lipid peroxidation and oxidative damage to DNA. In *Lipid Peroxides in Biology and Medicine*, K. Yagi, ed., Academic, New York, 1982, pp. 339–371.
3. B. N. Ames, and R. L. Saul. Oxidative DNA damage, cancer, and aging. In *Oxygen Radicals in Human Disease*, C. E. Cross, moderator, pp. 536–539. *Ann. Intern. Med.* **107**, 526–545 (1987).
4. A. Bakka, A. S. Johnsen, L. Endresen, and H. E. Rugstad. Radioresistance in cells with a high content of metallothionein. *Experentia* **38**, 381–383 (1982).
5. B. H. J. Bielski. Fast kinetic studies of dioxygen-derived species and their metal complexes. *Phil. Trans. Roy. Soc. Lond. B* **311**, 473–482 (1985).
6. W. Bohr, D. H. Phillips, and P. C. Hanawalt. Heterogenous DNA damage and repair in the mammalian genome. *Cancer Res.* **47**, 6426–6436 (1987).
7. E. T. Borish, and W. A. Pryor. Cigarette smoking, free radicals, and free radical DNA damage, pp. 535–536. In *Oxygen Radicals and Human Disease*, C. E. Cross, moderator. *Ann. Intern. Med.* 107, 526–545 (1988).
8. K. Brawn, and I. Fridovich. Superoxide radical and superoxide dismutases: threat and defense. *Acta Physiol. Scand. Suppl.* **482**, 9–18 (1980).
9. O. Cantoni, R. M. Evans, and M. Costa. Similarity in the acute cytotoxic response of mammalian cells to mercury (11) and X-rays: DNA damage and glutathione depletion. *Biochem. Biophys. Res. Comm.* **108**, 614–620 (1982).
10. P. A. Cerutti. Prooxidant status and tumor promotion. *Science* **227**, 375–381 (1985).
11. P. Cerutti. Oxidant tumor promoters. In *Growth Factors, Tumor Promoters and Cancer Genes*, Alan R. Liss, New York, 1988, pp. 239–247.
12. H. Chan, and W-S. Oxygen free radicals in food. *Proc. Nutr. Soc.* **46**, 35–41 (1987).
13. M. G. Cherian, and M. Nordberg. Cellular adaptation in metal toxicity and metallothionein. *Toxicology* **28**, 1–15 (1983).
14. R. Chiu, R. Imbra, M. Imagawa, and M. Karin. Metallothionein structure and function in regulating the trace element in human. In *Essential and Toxic Trace Elements in Human Health and Disease*, A. S. Prasad, ed., Alan R. Liss, New York, 1988, pp. 393–406.
15. N. T. Christie, and M. Costa. In vitro assessment of the toxicity of metal compounds III. Effects of metals on DNA structure and function in intact cells. *Biol. Trace Element Res.* **5**, 57–71 (1983).
16. N. T. Christie, and M. Costa. In vitro assessment of the toxicity of metal compounds IV. Disposition of metals in cells: interactions with membranes,

glutathione, metallothionein, and DNA. *Biol. Trace Element Res.* **6**, 139–158 (1984).

17. C. E. Cochrane, I. U. Schraufstater, P. Hyslop, and J. Jackson. Cellular and biochemical events in oxidative injury. In *Oxygen Radicals and Tissue Injury: Proceedings of an Upjohn Symposium,* B. Halliwell, ed., Federation of American Societies for Experimental Biology, Bethesda, 1988, pp. 49–54.

18. D. E. Coppen, D. E. Richardson, R. J. Cousins. Zinc suppression of free radicals induced in cultures of rat hepatocytes by iron, *t*-butyl hydroperoxide, and 3-methylindole. *Proc. Soc. Exp. Biol. Med.* **189**, 100–109 (1988).

19. P. I. Countryman, and J. A. Heddle. The production of micronuclei from chromosome aberrations in irradiated cultures of human lymphocytes. *Mutation Res.* **41**, 321–332 (1976).

20. P. M. Cullis, and M. C. R. Symons. Electron spin resonance studies of the mechanism of radiation damage to DNA. In *Mechanisms of DNA Damage and Repair,* M. G. Simic, L. Grossman, and A. C. Upton, eds., Plenum, New York, 1980, pp. 29–37.

21. J. Darnell, H. Lodish, and D. Baltimore. Cancer. In *Molecular Cell Biology,* Scientific American Books, New York, 1986, pp. 1035–1081.

22. P. B. Dervan. Design of sequence-specific DNA binding molecules. *Science* **25**, 464–471 (1986).

23. I. E. Dreosti, E. J. Partick, and I. R. Record. Micronutrients and micronuclei in X-irradiated mouse splenocytes. *Teratology,* **40**, 261 (1989).

24. M. Fenech, and A. A. Morley. Measurement of micronuclei in lymphocytes. *Mutation Res.* **147**, 29–36 (1985).

25. A. C. M. Filho, M. E. Hoffman, and R. Meneghini. Cell killing and DNA damage by hydrogen peroxide are mediated by intracellular iron. *Biochem. J.* **218**, 273–275 (1984).

26. G. L. Floersheim, and P. Floersheim. Protection against ionizing radiation and synergism with thiols by zinc aspartate. *Br. J. Radiol.* **59**, 597–602 (1986).

27. R. A. Floyd. DNA-ferrous iron catalyzed hydroxyl free radical formation from hydrogen peroxide. *Biochem. Biophys. Res. Comm.* **99**, 1209–1215 (1981).

28. B. A. Freeman, and J. D. Crapo. Biology of disease: Free radicals and tissue injury. *Lab. Invest.* **47**, 412–426 (1982).

29. A. W. Girotti, J. P. Thomas, and J. E. Jordan. Inhibitory effect of zinc (11) on free radical lipid peroxidation in erythrocyte membranes. *J. Free Rad. Biol. Med.* **1**, 395–401 (1985).

30. M. H. N. Golden, and D. Ramadath. Free radicals in the pathogenesis of Kwashiorkor. *Proc. Nutr. Soc.* **46**, 53–68 (1987).

31. J. M. C. Gutteridge, and X-C. Fu. Enhancement of bleomycin-iron free radical damage to DNA by antioxidants and their inhibition of lipid peroxidation. *FEBS Lett.* **123**, 71–74 (1981).

32. J. M. C. Gutteridge, and X-C. Fu. Protection of iron-catalyzed free radical damage to DNA and lipids by copper (11) bleomycin. *Biochem. Biophys. Res. Comm.* **99**, 1354–1360 (1981).

33. B. Halliwell. Free radicals and metal ions in health and disease. *Proc. Nutr. Soc.* **46**, 13–26 (1987).

34. B. Halliwell. Oxidants and human disease. Some new concepts. *FASEB J.* **1**, 358–364 (1987).

35. B. Halliwell, and J. M. C. Gutteridge. Oxygen toxicity, oxygen radicals, transition metals, and disease. *Biochem. J.* **219**, 1–14 (1984).

36. H. A. O. Hill. Oxygen, oxidases, and essential trace metals. *Phil. Trans. R. Soc. Lond.* **B294**, 119–126 (1981).

37. P. E. Hunziker, and J. H. R. Kägi. Metallothionein: a multigene protein. In *Essential and Toxic Trace Elements in Human Health and Disease*, A. S. Prasad, ed., Alan R. Liss, New York, 1988, pp. 349–363.
38. N. Imura, A. Naganuma, M. Satoh, and J-T. Chen. Trace elements as useful tools for cancer chemotherapy. In *Essential and Toxic Trace Elements in Human Health and Disease*, A. S. Prasad, ed., Alan R. Liss, New York, 1988, pp. 443–456.
39. S. Inouye. Site specific cleavage of double-strand DNA by hydroperoxide of linoleic acid. *FEBS Lett.* **172**, 231–234 (1984).
40. S. Inoue, and S. Kawanishi. Hydroxyl radical production and human DNA damage induced by ferric nitrolotriacetate and hydrogen peroxide. *Cancer Res.* **47**, 6522–6527 (1987).
41. U. Kvist, S. Kjellberg, L. Bjlorndahl, M. Hammar, and G. M. Roomans. Zinc in sperm chromatin and chromatin stability in fertile men and men in barren unions. *Scand. J. Urol. Nephrol.* **22**, 1–6 (1988).
42. A. Lione. Ionizing radiation and human reproduction. *Reprod. Toxicol.* **1**, 3–16 (1987).
43. L. J. Machlin, and A. Bendich. Free radical tissue damage: protective role of antioxidant nutrients. *FASEB J.* **1**, 441–445 (1987).
44. J. Matsubara, K. Ishioka, Y. Shibata, and K. Katoh. Risk analysis of multiple environmental factors: radiation, zinc, cadmium, and calcium. *Environ. Res.* **40**, 525–530 (1986).
45. *Mechanisms of DNA Damage and Repair*, M. G. Simic, L. Grossman, and A. C. Upton, eds., Plenum, New York, 1986.
46. R. Meneghini, and M. E. Hoffman. The damaging action of hydrogen peroxide on DNA of human fibroblasts is mediated by a nondialyzable compound. *Biochem. Biophys. Acta* **608**, 167–173 (1980).
47. B. D. Michael, S. Davies, K. D. Held, Ultrafast chemical repair of DNA single and double strand break precursors in irradiated V79 cells. In *Mechanisms of DNA Damage and Repair*, M. G. Simic, L. Grossman, and A. C. Upton, eds., Plenum Press, New York, 1986, pp. 89–100.
48. U. Moser. Uptake of ascorbic acid by leukocytes. In *Annals of the New York Academy of Sciences, Third Conference on Vitamin C*, J. J. Burns, J. K. Rivers, L. J. Machlin, eds., New York Academy of Sciences, New York, 1987, pp. 200–215.
49. J. R. F. Muindi, B. K. Sinha, L. Gianni, and C. E. Myers. Hydroxyl radical protection and DNA damage induced by anthracycline-iron complex. *FEBS Lett.* **172**, 226–230 (1984).
50. C. E. Myers, A. Katki, and E. Travis. Effect of tocopherol and selenium on defenses against reactive oxygen species and their effect on radiation sensitivity. In *Annals of the New York Academy of Sciences, Vitamin E, Biochemical, Hematological, and Clinical Aspects*, B. Lubin and L. Machlin, eds., New York Academy of Sciences, New York, 1982, pp. 419–425.
51. N. O. Nartey, D. Banerjee, and M. G. Cherian. Immunohistochemical localization of metallothionein in cell nucleus and cytoplasm of fetal human liver and kidney and its changes during development. *Pathology* **19**, 233–238 (1987).
52. R. Nery. *Cancer. An Enigma in Biology and Society*, The Charles Press, Philadelphia, 1986, pp. 350, 351.
53. T. Norseth. Metal carcinogenesis. In *Annals of the New York Academy of Sciences, Living in a Chemical World. Occupational and Environmental Significance of Industrial Carcinogens*, C. Maltoni and I. J. Selikoff, eds., The New York Academy of Sciences, New York, 1988, pp. 377–386.

54. T. Ochi. Effects of glutathione depletion and induction of metallothioneins on the cytotoxicity of an organic hydroperoxide in cultured mammalian cells. *Toxicology* **50**, 257–268 (1988).

55. P. Okunieff, and H. D. Suit. Toxicity, radiation sensitivity modification, and combined drug effects of ascorbic acid with misonidazole in vivo on FSall murine fibrosarcomas. *JNCI* **79**, 337–381 (1987).

56. A. Petro, and C. H. Hill. Response of hepatic metallothionein to iron administration. Lack of correlation with serum corticosterone. *Biol. Trace Element Res.* **14**, 255–263 (1987).

57. W. A. Pryor. Cancer and free radicals. *Basic Life Sci.* **39**, 45–57 (1985).

58. J. E. Repine, R. B. Fox, and E. M. Berger. Hydrogen peroxide kills *staphylococcus aureas* by reacting with staphylococcal iron to form hydroxyl radical. *J. Biol. Chem.* **256**, 7094–7096 (1981).

59. M. Rosalski, E. Kuziemska, and R. Wierzbicki. Content of mercury in chromatin and level of metallothionein proteins in kidneys and livers of rats. *Biochem. Pharmacol.* **30**, 2177–2178 (1981).

60. M. P. Rosin, B. P. Dunn, and H. F. Stich. Use of intermediate endpoints in quantitating the response of precancerous lesions to chemopreventive agents. *Can. J. Physiol. Pharmacol.* **65**, 483–487 (1987).

61. T. K. Shires. Iron-induced DNA damage and synthesis in isolated rat liver nuclei. *Biochem. J.* **205**, 321–329 (1982).

62. D. Shulte-Frohlinde. Comparison of mechanisms for DNA strand break formation by the direct and indirect effect of radiation. In *Mechanisms of DNA Damage and Repair*, M. G. Simic, L. Grossman, and A. C. Upton, eds., Plenum, New York, 1986, pp. 19–27.

63. E. G. Sideris, S. C. Charalambous, A. Tsolomyty, and N. Katsaros. Mutagenesis, carcinogenesis, and the metal elements—DNA interaction. *Prog. Clin. Biol. Res.* **259**, 13–25 (1988).

64. M. G. Simic. Introduction to mechanisms of DNA damage and repair. In *Mechanisms of DNA Damage and Repair*, M. G. Simic, L. Grossman, and A. C. Upton, eds., Plenum, New York, 1986, pp. 1–8.

65. M. G. Simic, and S. V. Javanovic. Free radical mechanisms of DNA base damage. In *Mechanisms of DNA Damage and Repair*, M. G. Simic, L. Grossman, and A. C. Upton, eds., Plenum, New York, 1986, pp. 39–49.

66. T. F. Slater, K. H. Cheeseman, M. J. Davies, K. Proudfoot, and W. Xin. Free radical mechanisms in relation to tissue injury. *Proc. Nutr. Soc.* **46**, 1–12 (1987).

67. G. D. Smoluk, R. C. Fahey, P. M. Calabro-Jones, J. A. Aguilera, and J. F. Ward. Radioprotection of cells in culture by WR-2721 and derivatives: form of the drug responsible for protection. *Cancer Res.* **48**, 3641–3647 (1988).

68. H. F. Stich. Micronucleated, exfoliated cells as indicators for genotoxic damage and as markers in chemoprevention trials. *J. Nutr. Growth Cancer* **4**, 9–18 (1987).

69. P. J. Thornalley, and M. Vasak. Possible role for metallothionein in protection against radiation-induced oxidative stress. Kinetics and mechanisms of its reaction with superoxide and hydroxyl radicals. *Biochim. Biophys. Acta* **827**, 36–44 (1985).

70. C. von Sonntag. Peroxyl radicals of nucleic acids and their components. In *Mechanisms of DNA Damage and Repair*, M. G. Simic, L. Grossman, and A. C. Upton, eds., Plenum, New York, 1986, pp. 51–58.

71. M. W. Webber, S. M. Maseehur Rehman, and G. T. James. Metallothionein induction and deinduction in human prostatic carcinoma cells: relationships

with resistance and sensitivity to adriamycin. *Cancer Res.* **48,** 4503–4508 (1988).

72. U. Weser, and E. Bischoff. Incorporation of ^{65}Zn in rat liver nuclei. *Eur. J. Biochem.* **12,** 571–575 (1970).
73. R. J. P. Williams. Zinc: what is its role in biology? *Endeavour* **8,** 65–70 (1984).
74. R. J. P. Williams. The necessary and desirable production of radicals in biology. *Phil. Trans. R. Soc. Lond.* **B311,** 593–603 (1985).
75. R. L. Willson. Iron, zinc, free radicals, and oxygen in tissue disorders and cancer control. In *Iron Metabolism,* Ciba Foundation Symposium 51, Elsevier, North Holland, 1977, pp. 331–354.
76. R. L. Willson. Vitamin, selenium, zinc, and copper interactions in free radical protection against ill-placed iron. *Proc. Nutr. Soc.* **46,** 27–34 (1987).
77. F. Y. Wu, and C. W. Wu. Zinc in DNA replication and transcription. *Ann. Rev. Nutr.* **7,** 275–271 (1987).
78. R. Zimmerman, and P. Cerutti. Active oxygen acts as a promoter of transformation in mouse embryo C3H/10TA/C18 fibroblasts. *Proc. Natl. Acad. Sci. USA* **81,** 2085–2987.

From: *Trace Elements, Micronutrients, and Free Radicals* • Ed.: I. E. Dreosti • ©1991 The Humana Press Inc.

CHAPTER 8

The Role of Free Radicals in Cancer and Aging

T. Mark Florence

ABSTRACT

There is increasing evidence that the progress of biologic aging results from accumulated free radical damage, and that a significant fraction of cancer is also caused by deleterious free radical reactions. Free radicals originate from the utilization of oxygen and the metabolism of organic compounds, and can be scavenged in living organisms by a range of enzymes and small antioxidant molecules. In this review, the various theories of aging are examined, as are the possible origins of cancer. The mitochondrion is proposed as the common link between cancer and aging, and the free radical reactions that occur in the mitochondrion are explained. Some of the important free radical reactions, both essential and deleterious, that occur continuously in a living organism, are discussed, and an outline is given of the multi-layered defense system that all aerobic organisms have against excess free radicals. The possibility that rate of aging and cancer incidence can be reduced by dietary supplementation with antioxidants is discussed in light of human and animal experiments, and suggestions are given for future research into the role of free radicals in health and disease.

1. INTRODUCTION

It is surprising to realize that biogerontology, the science of the aging process, has been actively studied only in the past 15–20 y (17,58,61) Why a process that is such a basic part of life was neglected for so long is

171

difficult to understand. Even now, the sudden burgeoning of interest in aging research results more from economic than from humanitarian considerations. Governments around the world have realized that the care of an increasingly aged population will impose huge costs on future societies (26). No nation has ever had to face this problem before, because zero population growth and low mortality have never occurred simultaneously (17). By the turn of the century, in Western countries, one person in six will be over the age of 65 (17). Many of these people will have chronic illnesses and, unless there is some dramatic improvement in the general health of the aged, a high percentage will require nursing home care. Since most of the diseases suffered by them are associated with old age, an assault on the aging process itself is the most likely method for achieving an overall improvement in the health of the aged (17).

Perhaps a general belief that aging is an unavoidable, natural process of life led to the previous lack of interest in biogerontology. Yet, other "natural" diseases, such as smallpox, diphtheria, and poliomyelitis, have been conquered. The degenerative conditions that make up the aging process, and which cause so much suffering and loss of dignity, should be considered as just another serious disease to be attacked with all the weapons in the armory of science and medicine. In Huxley's Brave New World (63), the citizens of that future society lived to about 70 with little diminution in their health, activity, or virility, then died quietly within a week or two. This "square" survival curve may not be completely attainable, but it is a goal toward which a great deal of research should be directed (26).

A major improvement in the health of the aged is more likely to be brought about by preventive medicine than by ad hoc treatment of the degenerative diseases as they appear. Preventive medicine has been sadly neglected in Western society but, in future, there must be more emphasis on people helping themselves to stay healthy. To achieve this, medical practitioners must become health educators as well as repairmen. Surely the true art of medicine lies not in the ability to heal, but in the ability to prevent disease (26,79).

In contrast with biogerontology, cancer research has received enormous support, and is by far the most protracted, intensive, and expensive scientific research program ever undertaken by man. Yet, if the aim of cancer research is the elimination or control of the disease of cancer, then this research has been spectacularly unsuccessful (5,26,75). Considerable advances have been made in the treatment of some of the less common cancers, such as childhood leukemia, lymphomas, and choriocarcinoma, but both the incidence rates and the survival times for the main cancers, i.e., colorectal, lung, and breast cancers, have changed little in the last 30–40 y (5,26). For example, in Australia between 1950 and 1980, the survival rate of patients with colorectal cancer showed no improvement whatsoever (62). Similar results were reported from other countries (5).

Table 1
Cancer Rates in Japan and in Japanese Migrants 1968–1972

Site of cancer	Japan	Japanese in Hawaii[a]	Caucasians in Hawaii
Stomach	1311	397	217
Colon	83	371	363
Breast(F)	315	1221	1869
Prostate	14	154	343

[a]Third generation

Table 2
Factors Involved in Cancer

Factor	Estimated percentage of all cancer deaths
Diet	35
Tobacco smoking	30
Sexual behavior	7
Occupation	4
Pollution	2

Doll and Peto's authoritative review on the epidemiology of cancer (*19*), and various studies of the change in cancer incidence with migration (Table 1), and in isolated communities in the same country (*19*), prove beyond reasonable doubt that diet is the largest single factor in the cause of cancer (Table 2). There is increasing evidence from prospective epidemiological studies that specific dietary factors, such as cruciferous vegatables (*4*), selenium (*90,91,113*), vitamin E (*91,105*), vitamin C (*10,38*), and β-carotene (*3,105*), can have a profound protective influence against cancer. This raises the intriguing possibility of significantly lowering cancer rates by choice of diet, and the use of natural dietary supplements. The use of dietary supplements, plus the elimination of tobacco smoking, could, theoretically, reduce the incidence of cancer to less than one-half its present level (*19*) (Tables 1, 2).

It is disappointing that only a tiny fraction of the funds allocated each year to cancer research are directed toward prevention of this disease. It is obviously preferable, on both humanitarian and economic grounds, to prevent rather than attempt to cure, such an intractable condition.

2. THE CAUSES OF AGING AND CANCER

2.1. Aging

Hayflick (*58*) has classified the main theories of aging under the headings of organ theories, physiological theories, and genome-based theories.

2.1.1. Organ Theories

2.1.1.1. IMMUNOLOGY   With age, the ability of the immune system to produce antibodies diminishes, with the thymus-dependent system showing the greatest decline. In addition, autoimmune manifestations increase greatly with age, resulting in an increase in autoimmune diseases. It is possible, however, that age associated diseases themselves cause the observed immunological deficits, hence, it is unproven that changes in the immune function are the fundamental cause of aging.

2.1.1.2. NEUROENDOCRINOLOGY   The loss of neurons and endocrine cells with age has been recognized for some time, and perturbations in these systems would have important and widespread effects on the body. For example, steroid loss at menopause may result in osteoporosis, and decrements in the hypothalmus would have a cascading effect on many endocrine target cell functions. However, the neuroendocrine theory lacks universal application, since not all organisms that age have complex neuroendocrine systems.

2.1.2. Physiological Theories

2.1.2.1. FREE RADICAL THEORIES   Free radicals are atoms or molecules having one or more unpaired electrons (Section 3). Because nature prefers electrons to be paired, free radicals are usually very unstable, and will react rapidly with other chemical species to obtain the additional electron. These free radical reactions can cause damage to genetic material (DNA), or lead to oxidation of membrane lipid material (Sections 3–6). Harman (53–57) has speculated that the process of biologic aging is simply the sum of all the deleterious free radical reactions occurring throughout the body. The mitochondrion may be the prime site of free radical reactions that lead to cancer and aging (Section 6).

Although more evidence needs to be accumulated about the biological effects of free radicals, the free radical theory of aging is attractive because of its universality and the increasing experimental data indicating the role of free radicals in the life process (6).

2.1.2.2. CROSSLINKAGE THEORY   This theory proposes that age changes result when two or more macromolecules become linked covalently or by a hydrogen bond. These linkages, which are irreversible, accumulate with time, and lead to inert or malfunctioning molecules. Because free radicals are efficient crosslinking agents, the crosslinkage theory may simply be a corrollary of the free radical theory.

2.1.2.3. WASTE PRODUCT ACCUMULATION   Fluorescent pigments found in nondividing cells, such as neurons and skeletal and cardiac muscle cells, are referred to as lipofuscin. Lipofuscin is formed from the auto-oxidation of lipid material, which involves free radical chain reactions. The lipofuscin accumulates with age in cells to the point where it may impede cell function. In a 90 y old, as much as 7% of the intracellular volume may be occupied by lipofuscin granules. There is no direct evidence, however, that the accumulation of these waste products actually damages cells.

2.1.3. Genome-Based Theories

2.1.3.1. THE SOMATIC MUTATION AND ERROR ACCUMULATION THEORIES   The central concept in the somatic mutation theory is that the accumulation of sufficient mutations in somatic cells will produce physiological damage characteristic of aging. Experimentation to prove this theory has, however, been inconclusive.

Error accumulation in reiterated DNA sequences could lead to inaccuracy in protein synthesis, producing age-associated decrements in cell function. Again, results have been ambiguous.

Variation in the effectiveness of DNA repair mechanisms has also been proposed as an explanation of the rate of aging. With some selected species, life-span correlates with efficiency of DNA repair, but there are several important exceptions.

2.1.3.2. PROGRAMMED THEORY OF AGING   Advocates of this theory, the "pre-ordained" theory of aging, propose that a prescribed series of events is included in the genome, and that this leads to programmed aging, much as other messages written into the genome lead to the orderly expression of development. The theory has the virtue of simplicity, but it has no proof experimentally, and no explanation has been given as to how programmed aging is expressed in terms of deleterious cell changes.

2.2. Cancer

Chemicals, of one type or another, are believed to be the main cause of cancer (3,26,73,84). A large number of chemicals, from azo dyes to polycyclic hydrocarbons, are now known to induce cancer in a variety of organs. Molecular configuration of the substance has a profound effect on carcinogenicity; addition of a methyl group, or a slight shift in the position of a substituent group on a molecule, can increase or decrease carcinogenicity dramatically (73).

Carcinogens react with genetic material either directly (e.g., ethylene oxide) or indirectly after metabolic activation (e.g., benzo(a)-pyrene) to cause a somatic mutation (Fig. 1). The initiated cell, on division, produces altered (malignant) cells that are the start of a malignant tumor. The target for the carcinogen may be either nuclear or mitochondrial DNA (Section 6).

Some workers believe that hormones may also be causative agents in cancer, since it is possible to produce neoplasms of the pituitary and

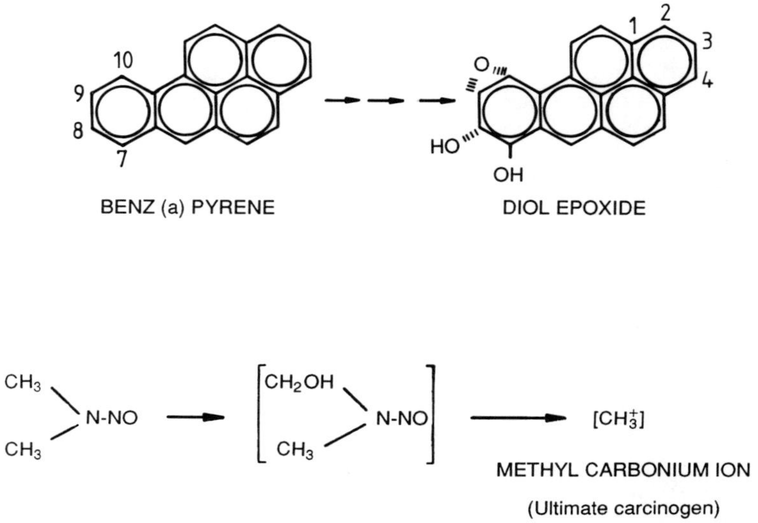

BENZ (a) PYRENE   DIOL EPOXIDE

METHYL CARBONIUM ION

(Ultimate carcinogen)

Fig. 1.   Metabolic activation of benzo(a)pyrene and dimethylnitrosamine to ultimate carcinogens.

endocrine organs in rodents by the administration of specific hormones (*84*). Chronic treatment with estrogen can induce mammary carcinomas in rodents, although this does not appear to be the case with humans. Evidence for the participation of a virus in rodent breast cancer has also been obtained, but no such virus has yet been identified in humans (*84*). A basic problem in cancer research is that the etiology of cancer is highly dependent on the species of the experimental animal.

There is indisputable evidence that most cancers are of environmental origin (*19,26*). Doll and Peto (*19*) estimated that 79–95% of all U.S. lung cancer, and at least 30% of all cancer deaths, are caused by smoking (Table 2). Skin cancer results principally from exposure to sunlight, the risk increasing with ultraviolet flux and less skin pigmentation. Australians and New Zealanders are particularly at risk, because people of Celtic origin (reddish complexion) are the most susceptible to skin cancer (*2,26*). The increased prevalence of lung and skin cancers is a good example of how fashion can influence the incidence of a disease. The habit of tobacco smoking, which became popular in the 1920s and 1930s, led to the epidemic of lung cancer in the 50s and 60s, whereas the myth, perpetuated by Hollywood, that a suntan is synonymous with health and success, has resulted in Australians having the world's highest rate of skin cancer (*2,26*).

Doll and Peto (*19*) showed conclusively that diet is the single greatest cause of cancer, followed by tobacco smoking (Table 1). Recent research (*67*) suggests the the contribution of diet to cancer may be even higher. Cancer caused by diet may depend not only on the presence of specific

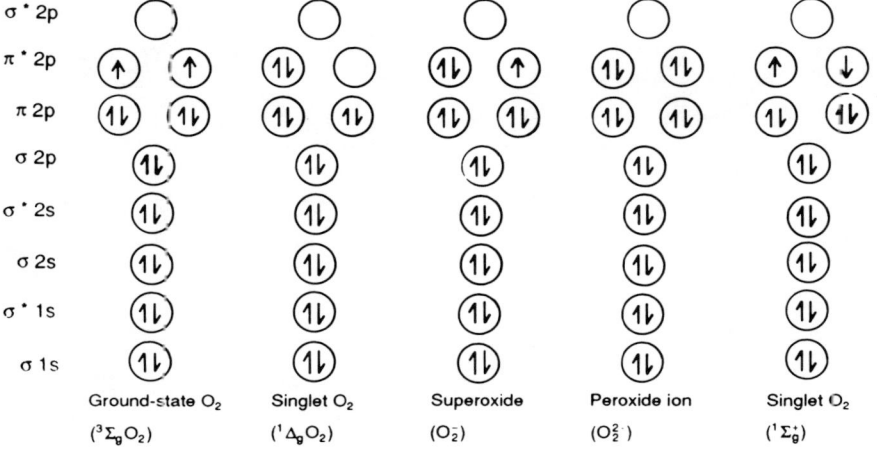

Fig. 2.   Electron distributions in some oxygen species. From reference 47, with permission.

carcinogens in food, but also by the lack of protective substances (e.g., free radical scavengers, Section 4) in it (2,26).

Epstein (22) argued strongly that occupation and industrial chemicals cause a substantial proportion of U.S. cancer, although Doll and Peto's (19) study (Table 2) showed that occupation and industrial products accounted for only about 5% of all cancer deaths. However, relatively few workers are exposed to industrial carcinogens, and the 5% may translate to quite a high occupational cancer rate in those exposed workers.

3. THE NATURE OF FREE RADICALS

3.1. Introduction

A free radical is any atom or molecule that contains one or more unpaired electrons; an unpaired electron being one that occupies an atomic or molecular orbital on its own. Figure 2 shows the π^* 2p (outer) orbitals of some oxygen species (46). Even ground state molecular oxygen (the type we breath) is a free radical, which explains its high reactivity (oxidation) with many compounds (47). However, oxygen cannot normally accept a pair of electrons from a nonradical molecule during oxidation because spin reversal would need to occur before the vacant spaces in the π^* orbital could be filled. Thus, oxygen must accept electrons one at a time, which makes the kinetics of many of its reactions slow, unless it gains two single electrons from another free radical (46,85). By contrast, singlet oxygen ($^1\Delta_g O_2$) can readily accept a pair of

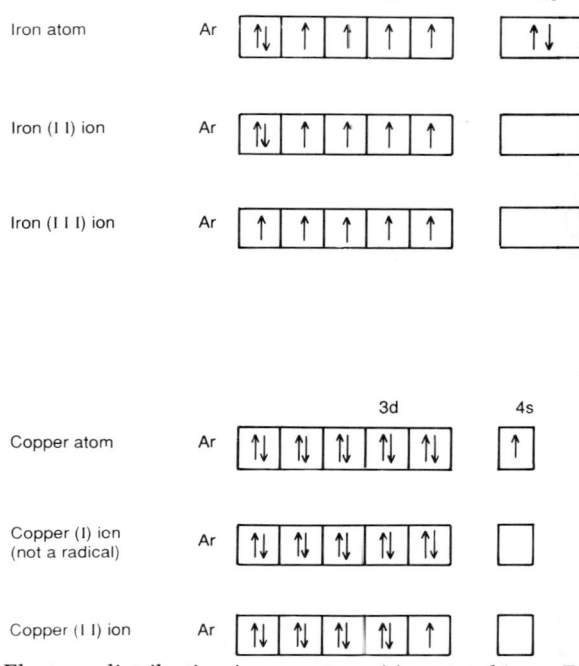

Fig. 3.   Electron distribution in some transition metal ions. From reference 47, with permission.

electrons, and its reactions with other molecules are very fast. Note that the peroxide ion is not a radical.

Most transition metals are free radicals (43,46) (zinc is an exception) (Fig. 3). Many transition metals can engage in one-electron valency state changes, which makes them excellent redox catalysts, e.g., Fe (II) − (III), Cu (I) − (II), and Mn(II) − (III). Some oxidases use these transition metals as catalysts at their active centers to overcome the sluggish reactions of molecular oxygen (47).

Table 3 lists some free radicals and activated oxygen species that have been implicated in biological processes, plus their approximate half-lives in a biological system (55,86,87). Some free radicals and activated oxygen species, e.g., hydrogen peroxide, lipid peroxide, and the semi-quinone free radical, are sufficiently stable to diffuse some distance in cells, whereas others, such as the hydroxyl radical (OH·), will react with the first organic molecule they encounter (86).

Other biologically important free radicals, not listed in Table 3, are nitrogen dioxide and several carbon-centered free radicals. Ozone (O_3), even though it is not a free radical, is a more powerful oxidizing agent than ground state oxygen. However, like many highly reactive nonradical species, most of the reactions of ozone with organic compounds involve free radical production (86,87).

Table 3
Approximate Half-Lives of Some Biological Free Radicals

Species	Symbol	Half-life(s) at 37°C[a]
Molecular oxygen	O_2	$>10^2$
Hydroxyl radical	OH·	1×10^{-9}
Superoxide radical	O_2^-	1×10^{-6}
Singlet oxygen	$^1O_2$	1×10^{-6}
Hydrogen peroxide	H_2O_2	10^b
Lipid peroxide	ROOH	$>10^2$
Alkoxyl radical	RO	1×10^{-6}
Peroxyl radical	ROO·	1×10^{-2}
Semiquinone radical	Q·⁻	$>10^2$

[a]In biological system.
[b]Short lifetime in presence of catalase or glutathione peroxidase.

Photochemical smog contains ozone and NOx (NO + NO_2). Both NO and NO_2 are stable free radicals that react rapidly with biological compounds. such as thiols and hemoglobin. Whereas smog usually contains less than lppm of NOx, undiluted cigarette smoke has several hundred ppm; in fact, gas-phase tobacco smoke contains 10^{17} reactive oxy radicals per puff (15,60)!

Carbon-centered free radicals are also common in biological systems. Here, the odd electron is located on a carbon, rather than an oxygen, atom. Similarly, nitrogen- (e.g., amines) and sulfur- (e.g., thiols) centered free radicals are commonly encountered.

3.2. Lipid Peroxidation

Lipid peroxidation is one of the most important free radical-mediated biological processes, and involves both oxygen- and carbon-centered free radicals (43,45,48). Lipid peroxidation involves attack on a polyunsaturated fatty acid molecule (e.g., linoleic acid) in a biological membrane, leading to decreased membrane fluidity, increased non-specific membrane permeability ("leaky" membranes), and inactivation of some membrane-bound enzymes. Accelerated lipid peroxidation can result from lack of free radical scavengers in the membrane (e.g., vitamin E) or increased oxidative stress as a result of the intake of xenobiotic drugs (e.g., antimalarial drugs), or breathing air with a higher than normal concentration of oxygen (18,21). The aging process (Section 5) almost certainly involves accelerated lipid peroxidation, and lipid peroxides have been called the "ultimate" toxin because of their long biological lifetime and their highly damaging nature (18,21,97).

Lipid peroxidation is a chain reaction (Fig. 4). A hydroxyl radical initiates the process by abstracting a hydrogen atom from a polyunsaturated fatty acid (PUFA) side chain (LH) to form water and a carbon-centered free radical (L·). This carbon radical then undergoes an internal

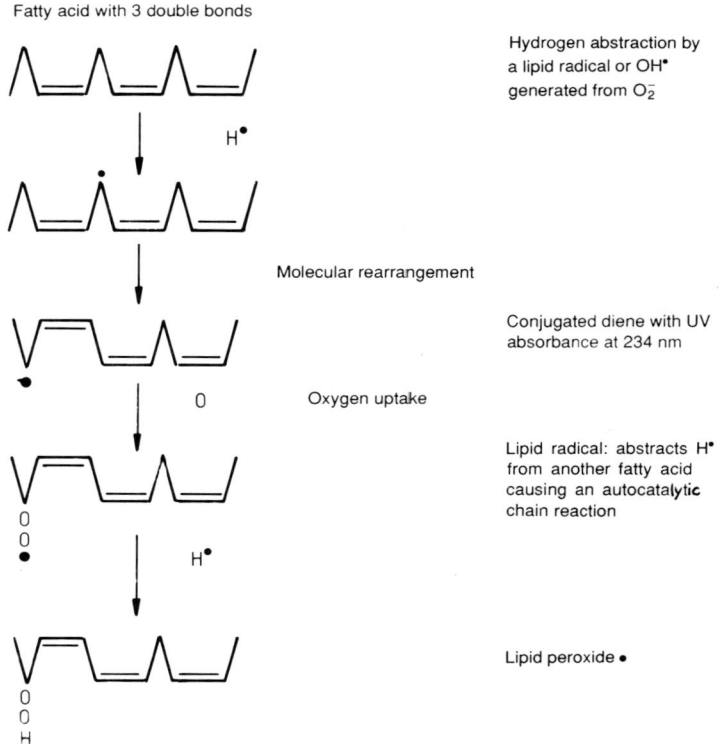

Fatty acid with 3 double bonds

Hydrogen abstraction by
a lipid radical or OH•
generated from O_2^-

Molecular rearrangement

Conjugated diene with UV
absorbance at 234 nm

O   Oxygen uptake

Lipid radical: abstracts H•
from another fatty acid
causing an autocatalytic
chain reaction

Lipid peroxide •

Fig. 4.   The chain reaction of lipid peroxidation.

rearrangement to yield a conjugated diene, that then reacts with molecular oxygen to give a peroxy radical (LOO·). The peroxy radical then abstracts a hydrogen atom from a second PUFA side chain to yield another L· and continue the chain reaction, forming a lipid peroxide (LOOH) in the process (*18,21*). A single hydroxyl radical can, thus, cause a calamitous cascade of membrane oxidation. The antioxidant vitamin E (α-tocopherol, EH) terminates this chain reaction by donating a hydrogen atom and trapping peroxy radicals (*107,109*).

$$LOO\cdot + EH \rightarrow LOOH + E\cdot \tag{1}$$

$$LOO\cdot + E\cdot \rightarrow \text{non-reactive product} \tag{2}$$

Vitamin E is the most efficient known terminator of lipid peroxidation; one molecule of α-tocopherol can protect 1,000 lipid molecules (*82*).

3.3. Source of Biological Free Radicals

Free radicals in biological systems can originate from a variety of sources. Molecular oxygen can react with a range of small organic molecules, including thiols, hydroquinones, flavins, and catecholamines, to yield superoxide radical (*8*).

$$O_2 + e^- \rightarrow O_2^{\bar{\cdot}} \qquad (3)$$

Hydrogen peroxide is a product of $O_2^{\bar{\cdot}}$ dismutation:

$$O_2 + O_2^{\bar{\cdot}} + 2H^+ \rightarrow H_2O_2 + O_2 \qquad (4)$$

Superoxide dismutase catalyzes this rather slow reaction, making it about 10^4 times faster (*46*).

Numerous enzymes generate superoxide radical during their catalytic cycling (*8,46*). These include xanthine oxidase, aldehyde oxidase, and dihydroorate dehydrogenase (*8*). Peroxisome enzymes, such as urate oxidase and D-amino-acid oxidase, are potent, direct sources of cellular H_2O_2, and the cytochrome P-450 (mixed function oxidase) system, responsible for the metabolism of foreign hydrophobic chemicals, produces both $O_2^{\bar{\cdot}}$ and H_2O_2 (*33,76*). Microsomal and plasma membrane-associated enzymes, such as lipoxygenase and cyclooxygenase, are utilized in arachidonic acid metabolism, a reaction that involves carbon-centered free radicals and a heme-bound hydroxyl radical (*33,76*).

The mitochondrial electron transport chain reduction of oxygen proceeds via oxygen free radicals and H_2O_2:

$$O_2 + e^- \rightarrow O_2^{\bar{\cdot}} \qquad (5)$$

$$O_2^{\bar{\cdot}} + e^- + 2H^+ \rightarrow H_2O_2 \qquad (6)$$

$$H_2O_2 + e^- + H^+ \rightarrow H_2O + OH\cdot \qquad (7)$$

$$OH\cdot + e^- + H^+ \rightarrow H_2O \qquad (8)$$

$$\text{Net: } O_2 + 4e^- + 4H^+ \rightarrow 2H_2O \qquad (9)$$

This reaction is catalyzed by the heme-copper enzyme cytochrome oxidase but, despite the high efficiency of this enzyme, some $O_2^{\bar{\cdot}}$ and H_2O_2 escape into the mitochondrion and the cytosol (*41*).

Hydroxyl radical can be produced directly as the result of the reaction between H_2O_2 and a suitable reductant, such as glutathione or NADH (*27,30*):

$$NADH + 2H_2O_2 + H^+ \rightarrow NAD^+ + 2H_2O + 2OH\cdot \qquad (10)$$

In addition, the iron- or copper-catalyzed Haber-Weiss reaction can also produce $OH\cdot$; $O_2^{\cdot-}$, and H_2O_2:

$$Fe(III) + O_2^{\cdot-} \rightarrow Fe(II) + O_2 \tag{11}$$

$$Fe(II) + H_2O_2 \rightarrow Fe(III) + OH^- + OH\cdot \tag{12}$$

Suitable catalysts for this fairly slow reaction are iron-EDTA and copper-1,10-phenanthroline (27). Vanadium, titanium, and cobalt redox couples can also act as catalysts (48). It is debatable, however, if effective iron and copper catalysts exist in vivo except in unusual diseases, since the presence of "free" (nonprotein bound) iron or copper in biological fluids has never been proven (15,27). Ferritin has been suggested (48,52) as a source of free iron, but iron may only dissociate from aged or degraded protein (27). However, by administering salicylate to volunteers, and measuring hydroxylated salicylate derivatives in urine, Halliwell et al. (50) provided some evidence that hydroxyl radical is produced in vivo. A hydroxyl radical bound to iron in a heme molecule ("crypto" hydroxyl radical) may be involved in biological oxidations by cytochrome-c and heme oxidases (28,48,112). An intriguing theory to explain the high rate of cardiovascular diseases in developed countries is based on the much higher iron status of people in affluent societies (108).

Singlet oxygen ($^1O_2$) is a potent oxidizer of lipid membranes (46,76). It can arise in biological systems from photochemical reactions involving porphyrins, flavins, and chlorophylls. It is also formed when hypochlorite reacts with H_2O_2 (51).

$$OCl^- + H_2O_2 \rightarrow Cl^- + H_2O + {}^1O_2 \tag{13}$$

Since hypochlorite is formed during phagocytosis with the enzyme myeloperoxidase, reaction (13) may be relevant in vivo (15,47).

Hydroxyl radical and singlet oxygen are the most reactive of the activated oxygen species. Their very reactivity, however, may limit the damage they can produce, since they can be scavenged so easily (15,26). For example, intracellularly-produced $OH\cdot$ was extremely toxic to the marine diatom *Nitzschia closterium*, whereas extracellularly-produced $OH\cdot$ was completely innocuous (29–31,100). Apparently, organic molecules in the growth medium or on the exterior of the cell membrane scavenged the hydroxyl radical before any damage could be done. The less reactive, but much more stable, H_2O_2 molecule, on the other hand, can cross cell membranes as readily as water, and could diffuse unimpeded to a vulnerable part of the cell where it could react with a suitable reductant to produce $OH\cdot$ (reaction 10), leading to damage of, for example, genetic material. Lipid peroxides are similarly very dangerous, because of their relative stability and selective reactivity (18,21).

It should be pointed out that oxygen free radicals, despite being so dangerous if generated in the wrong place or at too high a concentration, do perform several life-sustaining tasks in living organisms (*26,47*). They are essential to respiration (reactions 5–9) and they are the lethal agents in phagocytosis, where a circulating leucocyte engulfs a pathogenic microorganism and destroys it with a burst (the "respiratory burst") of O_2^-, H_2O_2, OCl^- and, possibly, $OH\cdot$ (*47*). Prostaglandin synthesis and the metabolism of alcohol also involve free radicals, and free radical modification of DNA may be necessary for heritable change (*102,104*).

4. FREE RADICAL SCAVENGERS

4.1. Utilization of Oxygen

All aerobic organisms have an impressive array of free radical scavengers. Many of these scavengers are designed specifically to protect the organism from oxygen-derived free radicals and other activated oxygen species. Man has the highest maximum lifespan potential of all mammals because he has superb free radical defenses (*17,80,102*).

Life apparently arose spontaneously 3.5 billion years ago from basic chemicals produced from the primitive oxygen-free atmosphere by free radical reactions initiated by ionizing radiation from the sun (*54*). About one billion years later, blue-green algae appeared, and some 1.3 billion years ago the concentration of atmospheric oxygen, produced by the photosynthesizing algae, had reached 1% of the present value, the toxic level for the fermentative anerobes (*41*). The anerobic procaryotes disappeared, except for a few in oxygen-deficient areas, and the sturdier, more complex, and more energy efficient eucaryotes became the dominant cells. Up to 18 times more energy in the form of ATP can be extracted from glucose by oxidizing it to CO_2, compared with anaerobic glycolysis (*41*).

The utilization of oxygen is, however, not without its problems. The in vivo reduction of oxygen produces O_2^-, $OH\cdot$, and H_2O_2 (reactions 5–9) that are highly damaging to the cell, and this toxicity increases rapidly if the oxygen concentration becomes much higher than the ambient 20% of the atmosphere. Our margin of safety is narrow. We possess defenses against oxygen toxicity that are sufficient to meet ordinary demands, but that can be easily overwhelmed (*46*).

4.2. Enzymatic Defenses

Superoxide dismutase (SOD) removes O_2^- via catalysis of reaction (4). There are three forms of SOD; copper–zinc, manganese, and iron (*17,78,103*). Copper–zinc and manganese SOD are found in eucaryotic cells (including human), whereas iron-SOD occurs in bacteria (*80,102*).

Copper–zinc SOD consists of two identical subunits, each having a single intramolecular S—S bond (*25*). These bonds are essential to the stability of the protein, and are unusually resistant to radiation and

DEFENSE LINE	TYPE OF DEFENSE	MAIN LOCATION
1st	CATALASE, Mn SOD	MT MATRIX
2nd	VIT E - MEMBRANE BOUND	MT INNER MEMBRANE
3rd	Cu,Zn SOD	MT INNER MEMBRANE SPACE AND CYTOPLASM
4th	GLUTATHIONE PEROXIDASE	CYTOPLASM
5th	ASCORBIC ACID, GLUTATHIONE, URIC ACID, CERULOPLASMIN, ETC.	SERUM, TISSUES, CYTOPLASM

MT = MITOCHONDRION

SOD = SUPEROXIDE DISMUTASE

Fig. 5.   Multilayered defense system in the cell.

chemical attack, a property doubtlessly essential to the efficient functioning of SOD as a free radical scavenger (25).

Two enzymes, glutathione peroxidase and catalase, are used to catalyze the decomposition of H_2O_2. Glutathione peroxidase has four atoms of selenium/mole, and uses glutathione as substrate to reduce H_2O_2 and lipid peroxides (93). Catalase is a hemoprotein, and has the advantage that it does not require an auxiliary reductant to destroy H_2O_2. These two enzymes, together with SOD, work in a synergistic fashion to protect lipid membranes and protein sulfhydryl groups, especially in the mitochondrion, from attack by O_2^- and H_2O_2 (42). Glutathione peroxidase also removes lipid peroxides and, thus, inhibits the chain reaction of lipid peroxidation (Fig. 4).

4.3. Nonenzymatic Defenses

The free radical dissociating enzyme defense system is backed up by an array of nonenzymatic defenses (a "strategic reserve"), consisting of small nucleophilic molecules that constitute a multilayered defense array against activated oxygen species, and which also scavenge carbon-centered free radicals (Fig. 5) (51,96,106). Some of these antioxidants, with approximate average concentrations in human blood plasma (mgL^{-1}) are: ascorbic acid (vitamin C), 10; reduced glutathione, 400 (whole blood); α-tocopherol (vitamin E), 10; uric acid, 50; β-carotene, 2; ceruloplasmin, 340 (3,79). Albumin and glucose also have free radical scavenging properties (51).

Vitamin E, like β-carotene, is lipid soluble, and occurs where it is

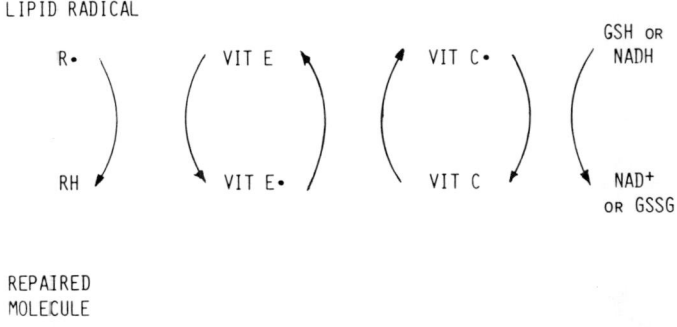

Fig. 6. Synergistic action of vitamins E and C and glutathione.

most needed, in the cell membrane, where it performs the critical task of terminating the potentially calamitous chain reaction of lipid peroxidation (Fig. 4). Ascorbic acid is not lipid soluble, but is a versatile reductant, and reacts synergistically with vitamin E and reduced glutathione (GSH) to produce a powerful antioxidant system (Fig. 6) (26,67,107).

4.4. Selenium

Selenium may be the most important antioxidant element in the human body (14,20,93). In addition to being essential to the functioning of glutathione peroxidase, it appears to have other more subtle roles, such as enhancing DNA repair mechanisms while delaying cell mitosis (64). This role may allow initiated cancer cells to repair themselves before division, so that the progeny are not malignant. An international study of the selenium content of the diet showed an excellent inverse correlation with cancer incidence (13), as did blood selenium and cancer incidence in Provinces across China (Fig. 7) (113). China is an ideal country to study the epidemiology of cancer and selenium, because the soil (and hence crops) in some areas of China is so low in selenium that a specific type of cardiomyopathy (Keshan Disease) occurs (93,113), whereas soil selenium is so high in other Provinces that chronic seleniosis sometimes occurs (113). New Zealanders are particularly low in selenium, and suffer high rates of cancer, cardiovascular disease, asthma, and sudden infant death syndrome (SIDS). Low dietary selenium has also been linked (88) with inflammatory diseases, such as rheumatism, arthritis, and repetitive strain injury, a not unlikely situation considering the free radical scavenging properties of selenium. A study has been initiated in Christchurch Hospital on the relationship between dietary selenium and SIDS (C. C. Winterbourn, private communication).

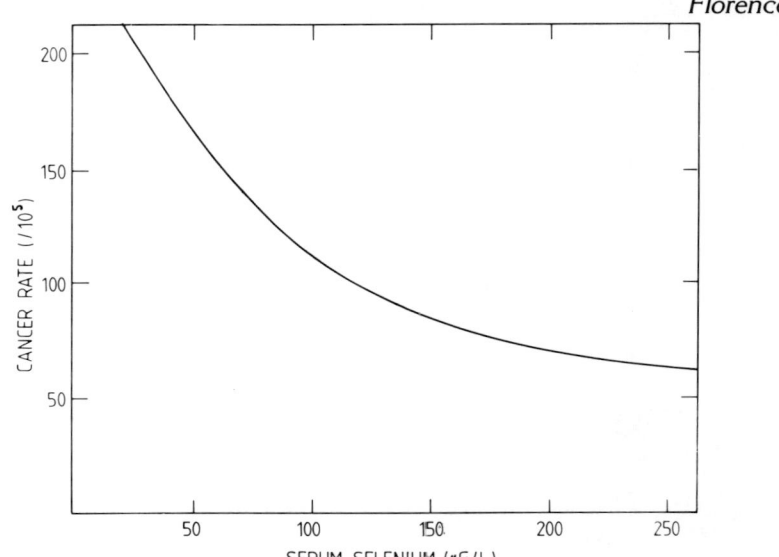

Fig. 7. The relationship between cancer rates (males) and blood selenium in China. Data from reference 113.

Selenium yeast is a cheap and convenient dietary selenium supplement for those who are not yeast sensitive.

5. INVOLVEMENT OF FREE RADICALS IN AGING

Various biological properties correlate with maximum life-span. For many mammals, the product of their average daily specific metabolic rate and maximum life-span is constant, at about 8×10^5 J/g (16,26,98). All mammals, whatever their size, breathe about 200 million times during their lives, whether that life be 2 or 80 y (40). These facts suggest that the mitochondrion, where most of the oxygen is processed, is the critical life-span site, and that only a certain "mileage" can be obtained from any mammalian mitochondrion. That is to say, the mitochondrion acts as a "biological clock" (53). A mouse has a life-span of only 2 y because its mitochondria are working 40 times as hard as those of a human.

Oxygen free radicals that escape the respiratory chain and normal scavenging processes will damage the mitochondrion membranes at a rate proportional to the rate of oxygen consumption, i.e., the metabolic rate. It has been estimated (26) that about 6% of the total oxygen uptake of a cell is "leaked" as O_2^-, which dissociates to H_2O_2 (reaction 4). The inner mitochondiral membrane, where these activated oxygen species are produced, may be the "Achilles Heel" of an aging cell (53,56). In

Table 4
Enzymes Known to Suffer an Age-Related Decline in Activity

Superoxide dismutase	Aldolase
Glutathione reductase	Maltase
Isocitrate lyase	3-Phosphoglycerate dehydrogenase
Pyruvate kinase	NADPH-cytochrome c
lactate dehydrogenase	(P-450) reductase
Phosphoglycerate kinase	Glutamine synthetase
Hypoxanthine/phosphoribosyltransferase	Enolase
Alcohol dehydrogenase	Aspartokinase III

mammalian cells, there is an age-related decrease in the number of mitochondria in fixed postmitotic cells (e.g., neurons and muscle cells), and in cells with low replication rates (23,24). Aged mitochondria are more fragile, have less DNA, produce less ATP, suffer a decrease in respiratory control, and lose their ability to divide, probably because of damage to the membranes and to mitochondrial DNA (74). In addition, aged mitochondria have lower rates of protein synthesis for the inner membrane (23,24,69).

Impaired mitochondrial function, with lower ATP production, would compromise the homeostasis of the organism, and lead to aging. Oberley et al. (77) suggested that damage by oxygen free radicals is the environmental factor that causes stem cells to lose their capacity to divide, and fully differentiated cells to die. They propose that this aging process is the natural mechanism for the replacement of old organisms with new, thus paving the way for natural selection and evolutionary change. Cancer initiation may result from free radical damage to the mitochondria of stem cells, whereas aging effects may occur mainly in postmitotic cells.

There is a gradual decline in aging cells in the GSH/GSSG and NADPH/NADP ratios (1,36,99). Since a decrease in the GSH/GSSG ratio is believed to lower the rate of cells mitosis (100), loss of GSH in an aged cell could be a reason for low rates of cell division.

In recent years, it has become clear that many proteins and enzymes are attacked by free radicals, and that some important enzymes become less active, by as much as 50%, as an animal ages (Table 4) (35, 36,39,69,79,89). The cytochrome P-450 mixed function oxidaze system is believed (35,79) to be the site of inactivation of these proteins. Methionine-sulfur is particularly susceptible to attack by O_2^-, $^1O_2$, and RO· to convert it to methionine sulfoxide (47). The gradual oxidative inactivation of proteins by free radicals may contribute to the aging process.

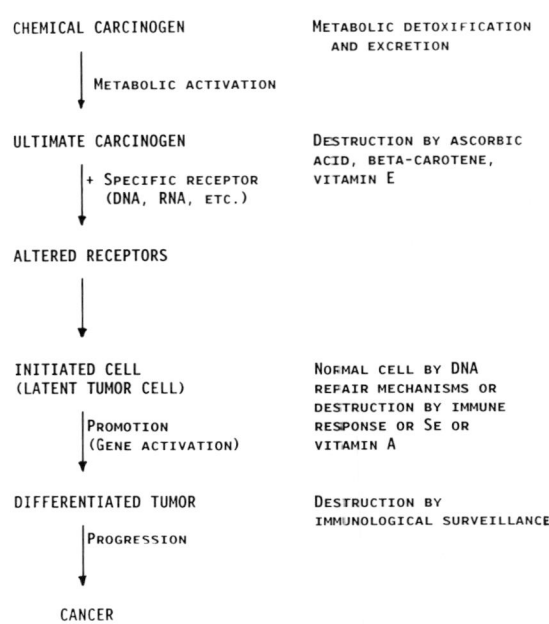

Fig. 8.   The progression of cancer and defenses against it.

6. INVOLVEMENT OF FREE RADICALS
IN CANCER AND OTHER DISEASES

6.1. Chemical Carcinogenesis

Cancer is a multistage process, involving both initiation and promotion steps (Fig. 8); free radicals are believed to be involved in both steps (*11,65,84*). Animal experiments have shown conclusively that dietary antioxidants (free radical scavengers) can intercede in both the initiation and promotion stages of cancer (*26,55,70*).

Most cancers appear to be initiated by environmental carcinogens, where "environment" includes dietary and atmospheric exposure (*19*). A wide range of natural and synthetic chemicals are known to be carcinogenic to animals (*3*). Some of these, as a result of industrial and social use, are known to be carcinogenic in man, whereas others are assumed to be so (Table 5). Man, with tobacco smoking, has carried out on his own species the largest cancer bioassay ever conducted.

Some chemicals are directly carcinogenic, but most require metabolic activation before they can react with genetic material. Activation involves metabolism by the cellular mixed-function oxidases, of which cytochrome P-450 is the terminal enzyme (*73,95*). Cytochrome P-450 is widely distributed in the cell, being found in the endoplasmic reticulum, the

Table 5
Some Chemicals Known or Suspected to Be Mutagenic in Man

Dietary Carcinogens	Occupational Carcinogens
Nitrosamines	
Safrole	2-Napthylamine
Cholesterol epoxide	Ethylene oxide
Methyleugenol	Benzo(a)pyrene
Psoralen derivatives	Dimethylbenzanthracene
Aflatoxin	Vinyl chloride
Quercetin	Methyleholanthrene
Quinones	Dimethyl sulfate
Vicine	Azo dyes
Malvalic acid	Ethyl carbonate
Theobromine	Ethionine
Gossypol	Mechlorethamine
	Nickel carbonyl

mitochondrion, and even in the nucleus itself (73,95). Miller and Miller (73) showed that activated carcinogens are positively-charged electrophilic compounds that react with nucleophilic centers in proteins or nucleic acids. Two main pathways are involved (73,95): hydroxylation at aliphatic or aromatic carbon atoms or heterocyclic rings, or epoxidation at aromatic or olefinic double bonds. Free radicals are probably involved in these activation reactions.

6.2. Cancer and the Mitochondrion

Most cancer may originate in the mitochondrion rather than in the cell nucleus. Mitochondria are the "powerhouses" of eucaryotic cells. Over 90% of the oxygen consumed by a mammal is utilized in the mitochondria, where pyruvate, fatty acids, carbohydrates, amino acids, and coenzymes are oxidized (53). This process produces CO_2 and water, and releases a large amount of energy, some of which is captured by the synthesis of ATP. Mitochondria are self-regulating, and contain their own DNA that directs the synthesis of some of the mitochondrial proteins (Fig. 9). Mitochondrial DNA is a single, circular DNA molecule, much less protected than the coiled and chromatin-packaged nuclear DNA (74). Mutagens bind to mitochondrial DNA up to 1,000 times more strongly than to nuclear DNA (24). Also, DNA repair mechanisms are much less efficient in the mitochondrion (24,110). Thus, both mitochondrial DNA and the organelle's inner and outer membranes (Fig. 9), which are high in polyunsaturated acids, are susceptible to attack by free radicals and electrophilic metabolites, despite the impressive multilayer free radical defense system (Fig. 5) (23,24,69).

Transformation of a cell from normal to malignant may result from damage to the mitochondrion. There is an as yet unexplained connection

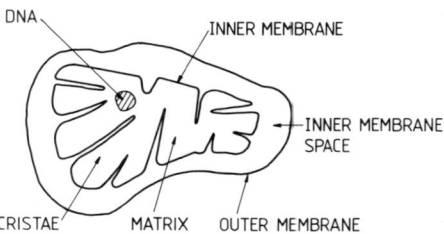

Fig. 9.   The mitochondrion.

between nuclear and mitochondrial DNA (perhaps even, direct exchange), and the inner mitochondrial membrane appears to direct the expression of cell surface antigens (*110*). Damage to the mitochondrion by oxygen free radicals leaking from the electron transport chain may cause a baseline level of cancer ("natural" cancer), whereas damage resulting from mutagenic metabolites of chemicals would account for the remainder (*103*).

Cancer cells contain less Mn-SOD than normal cells, and are less sensitive to oxygen (*78*). Cells transformed by chemical carcinogens will proliferate at oxygen concentrations fatal to normal cells. Immortality may be attained in cancer cells not only because they never lose their ability to divide, but because they are less susceptible to destruction by oxygen free radicals (*78*).

Damage to the mitochondria may explain the strong connection between cancer rates and age (*23,24*). Most cancer rates increase logarithmically with age, usually to the 4th or 5th power of age (*9*). It should be kept in mind, however, that although most cancer victims are elderly, initiation of cancer may take place 20 to 30 y before death (*9,84*). Cancer, then, is really a disease of the young and the middle-aged.

6.3. Other Free Radical Diseases

Free radicals are now believed to be involved in a wide range of human diseases (Table 6) (*15,33,37,44–51, 56,70,85,94,96*). The toxicity of many drugs, xenobiotics, such as quinones, nitroaromatics, and bipyridyls, and azo compounds, is caused by free radical formation resulting from the drugs undergoing redox reactions (*47*). Reperfusion injury, both in the heart and the gastrointestinal tract, is believed to result from free radical production when oxygenated blood is reintroduced to the ischemic tissue (*15,70,72*). The neurotoxic effects of manganese are the result of Mn(II)/(III) cycling, producing $O_2^{\cdot}$, H_2O_2, and OH· (*32*). Tobacco smoke contains a wide range of free radicals, both organic and inorganic, that are demonstrably mutagenic (*60,68*).

Table 6
Some Clinical Conditions in Which Free Radicals
Are Possibly Involved

Organ	Condition
Multiorgan	Autoimmune diseases
	Drug reactions
	Aging
	Cancer
	Radiation injury
Erythrocytes	Malaria
	Sickle cell anaemia
	Lad poisoning
Lung	Tobacco smoke effects
	Emphysema
	Hyperoxia
Cardiovascular system	Atherosclerosis
	Keshan Disease
Joint abnormalities	Rheumatoid arthritis
Brain	Parkinsons disease
	Senile dementias
Eye	Cataractogensis
	Retinopathy
Skin	Sking cancer
	Contact dermatitis

7. PROTECTION AGAINST CANCER AND AGING BY DIETARY CONTROL OF FREE RADICAL FORMATION

The possibility that cancer can be minimized, and the rate of aging reduced, by the intake of antioxidants (free radical scavengers) has intrigued nutritionists for the past 20 y (26,38,111). If cancer and aging are indeed caused by oxidizing free radicals, an increase in the concentration of antioxidant in tissues and the circulatory system may offer some protection. Certainly, many serum antioxidants show a decline with age (111), and an increase in dietary antioxidants usually brings about an increase in mean lifetime (but not maximum life-span) in experimental animals (55). Prospective human studies have shown that lower rates of cardiovascular disease and cancer are associated with a high status of serum vitamins A and E, β-carotene, and selenium (10,12,38,59,71,83, 91,92,101,105,109). Several epidemiological studies have indicated that the cruciferous vegetables, cabbage, broccoli, brussel sprouts, and cauliflower, protect against cancer (4,26). The active compound in these vegetables is believed to be a dithiolthione (4).

Based on the ascorbic acid content of a primitive vegetarian diet Pauling (81) estimated that modern man needs a diet with an average of 2.3 g of vitamin C/d. The degree of supplementation required will, of

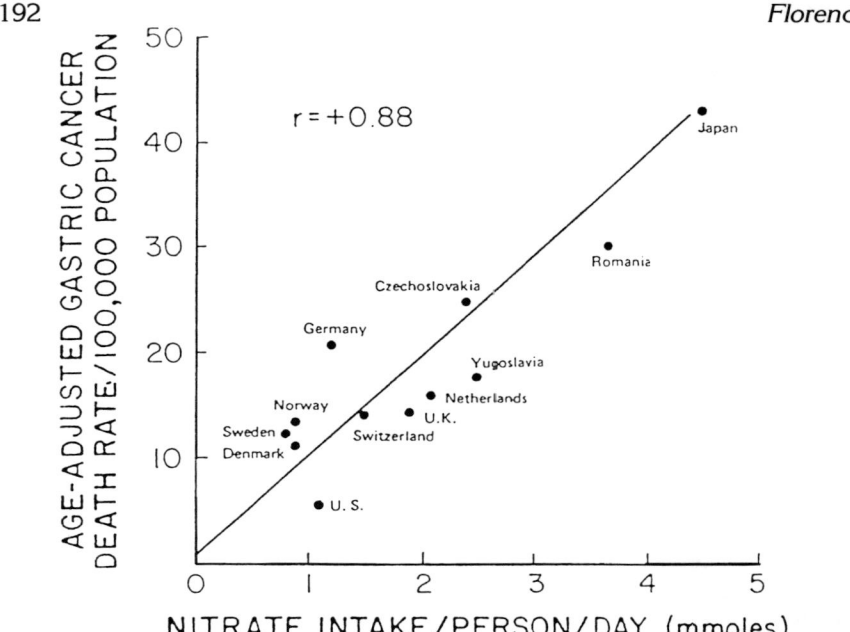

Fig. 10. International variation in gastric cancer with dietary intake of nitrate. From reference 92, with permission.

course, depend on individual diet and lifestyle, but 2.3 g of vitamin C/d will lead to an ascorbic acid serum concentration about three times the unsupplemented level (*81*). Tobacco smoking lowers serum vitamin C (each cigarette destroys about 25 mg of ascorbic acid), and some smokers suffer from chronic, subclinical scurvy (*82*).

Nitrosamines, formed from dietary amines and nitrite (present in preserved food or produced naturally from nitrate), are perhaps the most universal and potent class of carcinogen (*7,60,67,92*). We are exposed to them continuously, and they are probably the specific cause of gastric (stomach) cancer (*92*) (Fig. 10). Their formation is catalyzed by dietary compounds, such as thiocyanate, iodide, and polyphenols, and several nitrosamines are present in tobacco and tobacco smoke (*60*). Ascorbic acid destroys nitrosamines rapidly and completely, and a combination of vitamins C and E is even more effective (*67*). A high concentration of free vitamin C in the stomach, intestines, bladder, and tissues would seem to be desirable for protection against nitrosamines.

The argument is often put forward that intakes of vitamin C greater than about 150 mg per day are unnecessary because, when more than this is taken, vitamin C often appears in the urine. However, despite its appearance in the urine, much more than 150 mg/d is usually needed to ensure tissue saturation of the vitamin (*82*). There are no known ill effects of vitamin C when taken at the rate of 1 to 3 g/d (*82*). The frequently-raised connection between oxalate renal stones and vitamin C is false, and arose from inadequate analytical methods, in which urinary vitamin C interfered in the determination of oxalate (10,82).

Australia is one of the few countries in the world where the over-the-counter sale of dietary supplements containing selenium is prohibited. In excess, selenium, like many freely available drugs, such as aspirin, is toxic (93). But this is not sufficient reason for making it unavailable to the public. The Chinese cancer study (113) suggests (Fig. 7) that the optimum serum concentration of selenium is 250–300 µg/L. In most countries, supplementation, e.g., by low-cost selenium yeast tablets, would be necessary to achieve this serum concentration. If the tablets were prepared to contain, say, 50µg selenium/tablet, a huge excess over the recommended dosage would be needed to cause even low-grade chronic selenium toxicity (93).

8. FUTURE RESEARCH

The association between disease, especially age-related diseases, such as cancer, senile dementias, cardiovascular diseases, and arthritis and free radical formation is rapidly achieving more credence (15,70). Dietary antioxidants may offer protection against these free radical diseases, but research in this area is hindered by lack of funding and by skepticism from parts of the medical profession. With the escalating costs of medical treatment and of caring for an aging population, any simple preventive scheme, such as dietary supplementation, would have tremendous economic benefits. However, since deterioration from these age related diseases can take place over 30–50 y, supplementation should, ideally, be carried out for the whole of the adult life.

Ideally, all the vitamins and minerals needed for maximum protection from disease would be obtained from food alone. In a modern society, however, this is unlikely to occur because of variations in the nutrient content of food (e.g., soils are becoming deficient in selenium (34)), and because of the increasing addiction of most of the population to "fast" and packaged foods.

Research is urgently needed into the in vivo formation and detection of free radicals in humans (50), the existence in biological systems of factors, such as free iron and copper, that can catalyze the formation of hydroxyl radical, and the role of natural free radical scavengers. Previously unsuspected protective factors, such as germanium, which is present in garlic and onions (66), and the dithiolthiones in cruciferous vegetables (4), may be important. There may be many potent, naturally-occurring free radical scavengers that could profitably be used as dietary supplements. Large-scale prospective epidemiological studies on disease and antioxidants should be initiated to establish the role of dietary free radical scavengers, and to determine their optimum intakes for protection against the degenerative diseases.

Preventive medicine, sadly neglected in Western countries, must be taken much more seriously.

REFERENCES

1. W. A. Al-Turk, and S. J. Stohs. Hepatic glutathione content and aryl hydrocarbon hydroxylase activity of acetaminophen-treated mice as a function of age. *Drug Chem. Toxicol.* **4**, 37–48 (1981).
2. American Cancer Society. *Cancer Facts and Figures.* American Cancer Society, New York, 1986.
3. B. N. Ames. Dietary carcinogens and anticarcinogens. *Science* **221**, 1256–1264 (1983).
4. S. S. Ansher, P. Dolan, and E. Bueding. Biochemical effects of dithiolthiones. *Food Chem. Toxic.* **24**, 405–415 (1986).
5. J. C. Bailar, and E. M. Smith. Progress against cancer? *N. Engl. J. Med.* **314**, 1226–1232 (1986).
6. A. K. Balin, and R. G. Allen. Mechanisms of biologic aging. *Dermatol. Clin.* **4**, 347–358 (1986).
7. H. Bartsch, and R. Montesano. Relevance of nitrosamines to human cancer. *Carcinogenesis* **5**, 1381–1393 (1984).
8. J. Z. Byczkowski, and T. Gessner. Biological role of superoxide ion-radical. *Int. J. Biochem.* **20**, 569–580 (1988).
9. J. Cairns. *Cancer: Science and Society.* W. H. Freeman and Company, San Francisco, 1978.
10. R. F. Cathcart. Vitamin C: the nontoxic, nonrate-limited, antioxidant free radical scavenger. *Med. Hypotheses* **18**, 61–77 (1985).
11. P. A. Cerutti. Prooxidant states and tumor promotion. *Science* **227**, 375–381 (1985).
12. C. K. Chow. Nutritional influence on cellular antioxidant defense systems. *Amer. J. Clin. Nutr.* **32**, 1066–1081 (1979).
13. L. C. Clark. The epidemiology of selenium and cancer. *Fed. Proc.* **44**, 2584–2589 (1985).
14. L. C. Clark, and G. F. Combs. Selenium compounds and the prevention of cancer. Research needs and public implications. *J. Nutr.* **116**, 170–173 (1986).
15. C. E. Cross. Oxy radicals and human disease. *Ann. Intern. Med.* **107**, 526–545 (1987).
16. R. G. Cutler. Free radicals and aging. In *Molecular Basis of Ageing*, A. K. Roy and B. Chatterjee, eds., Academic, New York, 1984, pp. 263–354.
17. R. G. Cutler. Antioxidants, aging, and longevity. In *Free Radicals In Biology.* Vol. VI, W. A. Pryor, ed., Academic, New York, 1984a, pp. 371–428.
18. J. DiGuiseppi, and I. Fridovich. The toxicology of molecular oxygen. *CRC Crit. Rev. Toxicol.* **12**, 315–342 (1984).
19. R. Doll and R. Peto. The causes of cancer. Quantitative estimates of avoidable risks of cancer in the United States today. *J. Natl. Cancer Inst.* **66**, 1191–1308 (1981).
20. I. E. Dreosti. Selenium. *J. Food Nutr.* **43**, 60–78 (1986).
21. N. M. Emanuel. Kinetics and free-radical mechanisms of aging and carcinogenesis. In *Age Related Factors in Carcinogenesis*, A. Likhachev, V. Anisimov, and R. Montesano, eds., International Agency for Research on Cancer, Lyon, pp. 127–150.
22. S. Epstein. *The Politics of Cancer.* Anchor Press, New York, 1979.
23. J. E. Fleming, J. Miquel, and K. G. Bensch. Age dependent changes in mitochondria. *Basic Life Sci.* **35**, 143–156 (1985).
24. J. E. Fleming, J. Miquel, S. F. Cottrell, L. S. Yengoyan, and A. C. Economos. Is cell aging caused by respiration-dependent injury to the mitochondrial genome? *Gerontol.* **28**, 44–53 (1982).

25. T. M. Florence. Degradation of protein disulfide bands in dilute alkali. *Biochem. J.* **189,** 507–520 (1980).
26. T. M. Florence. Cancer and ageing. The free radical connection. *Chem. Australia* **50,** 166–174 (1983).
27. T. M. Florence. The production of hydroxyl radical from hydrogen peroxide. *J. Inorg. Biochem.* **22,** 221–230 (1984).
28. T. M. Florence. The degradation of cytochrome-*c* by hydrogen peroxide. *J. Inorg. Biochem.* **23,** 131–141 (1985).
29. T. M. Florence, J. L. Stauber, and K. J. Mann. The reaction of copper -2,9-dimethyl-1,10-phenanthroline with hydrogen peroxide. *J. Inorg. Biochem.* **24,** 243–254 (1985).
30. T. M. Florence. The production of hydroxyl radical from the reaction between hydrogen peroxide and NADH. *J. Inorg. Biochem.* **28,** 33–37 (1986).
31. T. M. Florence, and J. L. Stauber. Toxicity of copper complexes to the marine diatom *Nitzschia closterium. Aquatic Toxicol.* **8,** 11–26 (1986).
32. T. M. Florence, and J. L. Stauber. Manganese catalysis of dopamine oxidation. *Sci. Total Environ.* **78,** 223–240 (1989).
33. B. A. Freeman, and J. D. Crapo. Biology of disease: free radicals and tissue injury. *Lab. Invest.* **47,** 412–426 (1982).
34. D. V. Frost. What do losses in selenium and arsenic bioavailability signify for health? *Sci. Total Environ.* **28,** 455–466 (1983).
35. L. Fucci, C. N. Oliver, M. J. Coon, and E. R. Stadtman. Inactivation of key metabolic enzymes by mixed function oxidation reactions: Possible implication in protein turnover and ageing. *Proc. Natl. Acad. Sci. USA* **80,** 1521–1525 (1983).
36. A. Gafni. Age-related modifications in a muscle enzyme. In *Modifications of Proteins During Ageing,* R. C. Adelman and E. E. Dekker, eds., Alen R. Liss, New York, 1985, pp. 19–38.
37. K. F. Gey. On the antioxidant hypothesis with regard to arteriosclerosis. *Biblthca. Nutr. Dieta.* **37,** 53–91 (1986).
38. K. F. Gey, G. B. Brubacher, and H. B. Stähelin. Plasma levels of antioxidant vitamins in relation to ischemic heart disease and cancer. *Amer. J. Clin. Nutr.* **45,** 1368–1377 (1987).
39. G. A. Glass and D. Gershon. Enzymatic changes in rat erythrocytes with increasing cell and donor age. *Biochem. Biophys. Res. Comm.* **103,** 1245–1253 (1981).
40. S. J. Gould. One standard lifespan. *New Scientist* **81,** 388–389 (1979).
41. C. Greenwood, and H. A. Hill. Oxygen and life. *Chem. in Brit.* **18,** 194–196 (1982).
42. A. C. Griffin. Role of selenium in the chemoprevention of cancer. *Adv. Cancer Res.* **29,** 419–442 (1979).
43. J. M. Gutteridge, T. Westermarck, and B. Halliwell. Oxygen radical damage in biological systems. In *Free Radicals, Ageing, and Degenerative Diseases* J. E. Johnson, R. Walford, D. Harman, and J. Miquel, eds., Alan R. Liss New York, 1986, pp. 99–139.
44. B. Halliwell. Oxygen radicals: A commonsense look at their nature and medical importance. *Med. Biol.* **62,** 71–77 (1984).
45. B. Halliwell. Free radicals and metal ions in health and disease. *Proc. Nutr. Soc.* **46,** 13–26 (1987).
46. B. Halliwell and J. M. Gutteridge. Oxygen toxicity, oxygen radicals, transition metals, and disease. *Biochem. J.* **219,** 1–14 (1984).
47. B. Halliwell, and J. M. Gutteridge. *Free Radicals in Biology and Medicine.* Clarendon Press, Oxford, 1985.

48. B. Halliwell, and J. M. Gutteridge. The importance of free radicals and catalytic metal ions in human diseases. *Mol. Aspects Med.* **8**, 89–193 (1985).

49. B. Halliwell, and J. M. Gutteridge. Oxygen radicals and the nervous system. *Trends in Neurosci.* **8**, 22–26 (1985).

50. B. Halliwell, J. M. Gutteridge, and D. Blake. Metal ions and oxygen radical reactions in human inflammatory joint disease. *Phil. Trans. Royal Soc. Lond.* **B311**, 659–671 (1985).

51. B. Halliwell, and J. M. Gutteridge. Oxygen free radicals and iron in relation to biology and medicine: some problems and concepts. *Arch. Biochem. Biophys.* **246**, 501–514 (1986).

52. B. Halliwell, and J. M. Gutteridge. Iron and free radical reactions: two aspects of antioxidant protection. *TIBS* **11**, 372–375 (1986).

53. D. Harman. "The biologic clock: The mitochondria?" *J. Amer. Geriat. Soc.* **20**, 145–149 (1972).

54. D. Harman. The aging process. *Proc. Natl. Acad. Sci. USA* **78**, 7124–7128 (1981).

55. D. Harman. Free radicals and the origination, evolution, and present status of the free radical theory of aging. In *Free Radicals in Molecular Biology, Ageing, and Disease*, D. Armstrong, R. S. Sohal, R. G. Cutler, and T. F. Slater, eds., Raven, New York, 1984, pp. 1–42.

56. D. Harman. Free radical theory of aging: The "free radical" diseases. *Age 7*, 111–131 (1984).

57. D. Harman. Free radical theory of aging: Role of free radicals in the origination and evolution of life, aging, and disease processes. In *Free Radicals, Ageing, and Degenerative Diseases*, J. E. Johnson, R. Walford, D. Harman, and J. Miquel, eds., Alan R. Liss, New York, 1986, pp. 3–49.

58. L. Hayflick. Theories of biological aging. *Exp. Gerontol.* **20**, 145–159 (1985).

59. C. H. Hennekens. Micronutrients and cancer prevention. *N. Engl. J. Med.* **315**, 1288-1289 (1986).

60. D. Hoffmann. Chemical carcinogens in tobacco. In *Cancer Risks*, P. Bannasch, ed., Springer-Verlag, Berlin, 1987, pp. 95–113.

61. R. Holliday. The ageing process is a key problem in biomedical research. *Lancet* **(ii)**, 1386–1387 (1984).

62. E. S. Hughes, F. T. McDermott, A. L. Polglase, W. R. Johnson, and E. A. Pihl. Large bowel cancer—the next move? *Med. J. Australia* **1**, 36–37 (1982).

63. A. Huxley. *Brave New World.* Chatto and Windus, London, 1932.

64. C. Ip. Selenium inhibition of chemical carcinogenesis. *Fed. Proc.* **44**, 2573–2578 (1985).

65. T. W. Kensler, and M. A. Trush. Role of oxygen radicals in tumour promotion. *Environ. Mutagen.* **6**, 593–616 (1984).

66. P. M. Kidd. Germanium-132 (Ge-132): Homeostatic normalizer and immunostimulant. A review of its preventive and therapeutic efficacy. *Int. Clin. Nutr. Rev.* **7**, 11–20 (1987).

67. D. Lathia, A. Braasch, and V. Theissen. Inhibitory effects of vitamin C and E on in vitro formation of N-nitrosamine under physiological conditions. *Front. Gastrointest. Res.* **14**, 151–156 (1988).

68. L. A. Loeb, V. L. Emster, K. E. Wamer, J. Abbots, and J. Laszio. Smoking and lung cancer: An overview. *Cancer Res.* **44**, 5940–5958 (1984).

69. D. L. Marcus, N. G. Ibrahim, and M. L. Freedman. Age-related decline in the biosynthesis of mitochondrial inner membrane proteins. *Exp. Gerontol.* **17**, 333–341 (1982).

70. J. L. Marx. Oxygen free radicals linked to many diseases. *Science* **235**, 529–531 (1987).

71. H. R. Massie, V. R. Aiello, and T. J. Doherty. Dietary vitamin C improves the survival of mice. *Gerontology* **30,** 371–375 (1984).

72. J. M. McCord. Oxygen-derived free radicals in postischemic tissue injury. *N. Engl. J. Med.* **312,** 159–163 (1985).

73. E. C. Miller, and J. A. Miller. Searches for ultimate chemical carcinogens and their reactions with cellular macromolecules. *Cancer* **47,** 2327–2345 (1981).

74. J. Miquel, and J. Fleming. Theoretical and experimental support for an "oxygen radical—mitochondrial injury" hypothesis of cell aging. In *Free Radicals, Ageing, and Degenerative Diseases,* J. E. Johnson, R. Walford, D. Harman, and J. Miquel, eds., Alan R. Liss, New York, 1986, pp. 51–74.

75. C. S. Muir, and D. M. Parkin. The world cancer burden: prevent or perish. *Brit. Med. J.* **290,** 5–6 (1985).

76. A. Naqui, and B. Chance. Reactive oxygen intermediates in biochemistry. *Ann. Rev. Biochem.* **55,** 137–166 (1986).

77. L. W. Oberley, and T. D. Oberley. Free radicals, cancer, and aging. In *Free Radicals, Ageing and Degenerative Diseases,* J. E. Johnson, R. Walford, D. Harman, and J. Miquel, eds., Alan R. Liss, New York, 1986, pp. 325–371.

78. L. W. Oberley, T. D. Oberley, and G. R. Buettner. Cell differentiation, aging, and cancer. The possible roles of superoxide and superoxide dismutases. *Med. Hypotheses* **6,** 249–268 (1980).

79. C. N. Oliver, R. Fulks, R. L. Levine, L. Fucci, A. J. Rivett, J. E. Roseman, and E. R. Stadtman. Oxidative inactivation of key metabolic enzymes. In *Molecular Basis of Ageing,* A. K. Roy and B. Chatterjee, eds., Academic, New York, 1984, pp. 235–262.

80. T. Ono, and S. Okada. Unique increase of superoxide dismutase level in brains of long living mammals. *Exp. Gerontol.* **19,** 349–354 (1985).

81. L. Pauling. Evolution and the need for ascorbic acid. *Proc. Natl. Acad. Sci. USA* **67,** 1643–1648 (1970).

82. L. Pauling. *How to Live Longer and Feel Better.* Avon Books, New York, 1986.

83. R. Peto, R. Doll, J. D. Buckley, and M. G. Spron. Can dietary beta-carotene materially reduce human cancer rates? *Nature* **290** 201–208 (1981).

84. H. C. Pitot. *Fundamentals of Oncology.* Marcel Dekker, New York, 1981.

85. P. H. Proctor, and E. S. Reynolds. Free radicals and disease in man. *Physiol. Chem. Phys. Med. NMR* **16,** 175–195 (1984).

86. W. A. Pryor. Cancer and free radicals. *Basic Life Sci.* **39,** 45–59 (1986).

87. W. A. Pryor. Oxyradicals and related species: Their formation, lifetimes, and reactions. *Ann. Rev. Physiol.* **48,** 657–667 (1986).

88. M. F. Robinson. The New Zealand selenium experience. *Amer. J. Clin. Nutr.* **48,** 521–534 (1988).

89. M. Rothstein. The alteration of enzymes in aging animals. *Basic Life Sci.* **35,** 193–204 (1985).

90. J. T. Salonen, G. Alfthan, J. K. Huttunen, and P. Puska. Association between serum selenium and the risk of cancer. *Amer. J. Epidemiol.* **120,** 342–349 (1984).

91. J. T. Salonen, R. Salonen, R. Lappetelainen, P. H. Maenpaa, G. Alfthan, and P. Puska. Risk of cancer in relation to serum concentrations of selenium and vitamins A and E: matched case-control analysis of prospective data. *Brit. Med. J.* **290,** 417–420 (1985).

92. R. J. Shamberger. *Nutrition and Cancer,* Plenum, New York, 1984.

93. R. J. Shamberger. Selenium. In *Biochemistry of the Essential Ultratrace Elements,* E. Frieden, ed., Plenum New York, 1984, pp. 201–237.

94. D. M. Shankel, P. E. Hartman, T. Kada, and A. Hollaender. Synopsis of the first international conference on mutagenesis and anticarcinogenesis. *Environ. Mutagen.* **9**, 87–103 (1987).

95. P. Sims. Metabolic activation of chemical carcinogens. *Brit. Med. Bull.* **36**, 11–18 (1980).

96. T. F. Slater. Free radical mechanisms in tissue injury. *Biochem. J.* **222**, 1–15 (1984).

97. T. F. Slater, K. H. Cheeseman, and K. Proudfoot. Free radicals, lipid peroxidation, and cancer. In *Free Radicals in Molecular Biology, Ageing, and Disease*, D. Armstrong, R. S. Sohal, R. E. Cutler, and T. F. Slater, eds., Raven, New York, 1984, pp. 293–355.

98. R. S. Sohal, and R. G. Allen. Relationship between metabolic rate, free radicals, differentiation, and aging: A unified theory. *Basic Life Sci.* **35**, 75–104 (1985).

99. R. S. Sohal, P. L. Toy, K. J. Farmer. Age-related changes in the redox status of the housefly, *Musca domestica. Arch. Gerontol. Geriatr.* **6**, 95–100 (1987).

100. J. L. Stauber, and T. M. Florence. Mechanism of toxicity of ionic copper and copper complexes to algae. *Mar. Biol.* **94**, 511–519 (1987).

101. K. S. Sundaram, R. London, S. Manimerkalai, P. P. Nair, and P. Goldstein. α-Tocopherol and serum lipoproteins. *Lipids* **16**, 223–227 (1981).

102. J. M. Tomlasoff, T. Ono, and R. G. Cutler. Superoxide dismutase: correlation with life span and specific metabolic rate in primate species. *Proc. Natl. Acad. Sci. USA* **77**, 2777–2781 (1980).

103. J. R. Totter. Spontaneous cancer and its possible relationship to oxygen metabolism. *Proc. Natl. Acad. Sci. USA* **77**, 1763–1767 (1980).

104. M. Vuillaume. Reduced oxygen species, mutation, induction, and cancer initiation. *Mut. Res.* **186**, 43–72 (1987).

105. N. J. Wald, J. Boreham, J. L. Hayward, and R. D. Bulbrook. Plasma retinol, β-carotene, and vitamian E levels in relation to the future risk of breast cancer. *Brit. J. Cancer* **49**, 321–324 (1984).

106. D. D. Wayner, G. W. Burton, K. U. Ingold, L. R. Barclay, and S. J. Locke. The relative contribution of vitamin E, urate, ascorbate, and proteins to the total peroxyl radical-trapping antioxidant activity of human blood plasma. *Biochem. Biophys. Acta* **924**, 408–419 (1987).

107. H. Wefers, and H. Sies. The protection by ascorbate and glutathione against microsomal lipid peroxidation is dependent on vitamin E. *Europ. J. Biochem.* **174**, 353–357 (1988).

108. E. D. Weinberg. Iron witholding: A defense against infection and neoplasia. *Physiol. Rev.* **64**, 65–102 (1984).

109. R. L. Willson. Free radical protection: why vitamin E, not vitamin C, β-carotene or glutathione? Ciba Foundation Symposium No. 101, 1983, pp. 19–44.

110. M. Yaffe, and G. Schatz. The future of mitochondrial research. *TIBS* **9**, 179–181 (1984).

111. V. R. Young. Vitamins and the aging process. *Geriatrics* **2**, 418–435 (1984).

112. R. J. Youngman. Oxygen activation: is the hydroxyl radical always biologically relevant? *TIBS* **9**, 280–283 (1984).

113. S.-Y. Yu, Y.-J. Chu, X.-L. Gong, C. Hou, W.-G. Li, H.-M. Gong, and J.-R. Xie. Regional variation of cancer mortality incidence and its relation to selenium levels in China. *Biol. Trace Element Res.* **7**, 21–29 (1985).

From: *Trace Elements, Micronutrients, and Free Radicals* • Ed.: I. E. Dreosti • ©1991 The Humana Press Inc.

CHAPTER 9

Free Radicals and Malnutrition

MICHAEL H. N. GOLDEN, DAN D. RAMDATH, AND BARBARA E. GOLDEN

1. INTRODUCTION

A free radical is simply an atom or molecule with an unpaired electron. This confers very considerable reactivity on the molecule. As the unpaired electron seeks to be paired, it may either abstract an electron from a donor molecule, leaving a new radical in its place, or attach itself to a second molecule, forming an adduct: the position of the unpaired electron may then change to form a new radical. Only when two molecules react, both of which have unpaired electrons, will the radical reaction terminate without a radical product. The single electron step may be an intermediate in an essential biological process that incorporates a safe mechanism to provide the pairing electron. It is important to emphasize that these reactions are common to many of the normal transformations and energy producing metabolic steps of the body: free radicals are common and normal intermediates in metabolism. All the monooxygenases, several dehydrogenases, cytochrome-P_{450} and b_6, prostaglandin synthetase, leucotriene synthetase, vitamin K-dependent enzymes, the mitochondrial respiratory chain, and many other enzymes normally generate radicals. Photochemical production of radicals occurs on exposure to sunlight, and the production of radicals in the retina leads to changes in molecular configuration that are perceived as light. There is nothing esoteric or unusual about radical reactions: they are commonplace.

The body not only produces radicals during normal metabolism but it also purposefully produces radicals, designed to be toxic, during the immune and inflammatory responses. These radicals—oxygen and chlorine metabolites—are deliberately generated during the respiratory burst of the macrophage in order to kill invading organisms. They are thus an integral part of inflammation.

Radicals are thus "involved" in many disease processes just as surely as they are involved in normal metabolism. That is not the point. When someone is shot in a war, how do we apportion blame to the gun, the soldier, the general, the invader, and to the policymaker? We cannot, for they are all intimately involved. We can only progress by understanding the respective role of each component and its relationship to the others. In this respect, it is probably incorrect to talk about "free radical disease" *per se*. The bullets that do damage in many diseases are almost certainly free radicals and their products. They are currently of great interest because they provide an explanation for many steps in disease processes that we, as yet, do not understand and also, they are open to therapeutic intervention. In our analogy however, one would get a very mistaken picture of a war if we only looked at the number, caliber, and position of the guns. Attention has been focused, almost exclusively, on the generation of free radicals, particularly when excessive or in unwanted sites or situations. This is a part, but only a part, of the disease process. We should not fool ourselves into thinking that they are central to a disease process, with all other initiating and contributing factors subsidiary to them.

In considering the role of free radicals in disease, their generation is only part of the story: the other part is their control, containment, and safe disposal. Because radicals and their products are continuously being generated and are so reactive chemically, they must be closely controlled. They have to be passed from molecule to molecule in an orderly fashion without excess leakage. The reactions have to take place in parts of the cell and under circumstances in which they will not damage vital components. There must be an efficient protective and scavenging system for those radicals that do escape from their ordered path, and there must be an efficient system of repair and replacement of damage. If these components are defective, then free radical damage will occur. The protective systems are particularly important when excess radicals are purposely generated and released into the extracellular environment in order to kill invading organisms (or what the body perceives to be invading organisms in the form of antigen). The radicals produced are toxic for the cells of the body, just as they are for invading bacteria. There are mechanisms in the immune system for targeting the attack; nevertheless, it is clear that host tissues are damaged in all inflammatory reactions. What limits that damage is a complex protective and repair system, and a mechanism to dampen and terminate an inflammatory reaction.

Whether or not there is clinically relevant free radical induced tissue damage, either locally or generally, must depend on the **balance** between the **rate** of free radical production and the available **capacity** for their disposal. If the prooxidants overwhelm the antioxidants, then oxidative stress and damage will occur. Damage will not occur if: (1) the prooxidants are adequately constrained in activity and site, (2) cofactors required by the prooxidants (usually reducing equivalents) are unavailable, and (3) the appropriate antioxidants are present in adequate amounts at the site of generation. Those tissues that are normally exposed to a high flux of radicals are provided with correspondingly rich antioxidant defenses, at least in healthy individuals. The retina is an easily appreciated example—with photochemical radical production and abundant provision of antioxidants in health.

It is clear that if a proper assessment of the likelihood for free radical mediated damage is to be made, in any condition, then both the production of free radicals and the quality and quantity of the constraining, defense, and repair systems must be assessed. It is only when an imbalance between these exists that radicals are likely to be of clinical significance. This is true of all conditions in which free radicals have been implicated in the pathogenesis. In those conditions where radical production does not vary much and is quantifiable, such as radiation sickness and (radical producing) drug or toxin administration, the major determinant of the injury will be the adequacy, or otherwise, of the protective mechanisms. The quality of the protective machinery is largely dependent on the nutritional state of the subject.

There are at least fourteen different, essential nutrients involved in the various binding proteins, protective enzymes, coenzymes, regenerating pathways, and antioxidants that constitute the repertoire of antioxidant protection. The relationship of these nutrients is shown in Fig. 1, and their functions in Table 1. Perusal of this list shows just how extensive a range of nutrients the antioxidant protective system depends on. Clearly, the nutritional state of the patient will be a dominant factor in determining whether or not a given quantum production of "free" (uncontrolled) free radicals will, or will not, produce damage.

Each of the various protective systems, although individually important, and predominating in different domains, can, to some extent, compensate for each other. Thus, in experimental selenium deficiency, there is both an induction of glutathione S-transferase and compensatory protection from tocopherol. These can usually protect the person (animal). However, if the deficiency is acute, and accompanied by a relatively low intake of tocopherol, clear lesions develop, even with the normal rate of radical production. These animals are then vulnerable to additional stress.

Unlike those conditions characterized by a primary increase in radical production, the malnourished individual provides a model in which the primary problem is a decrease in the protective mechanisms.

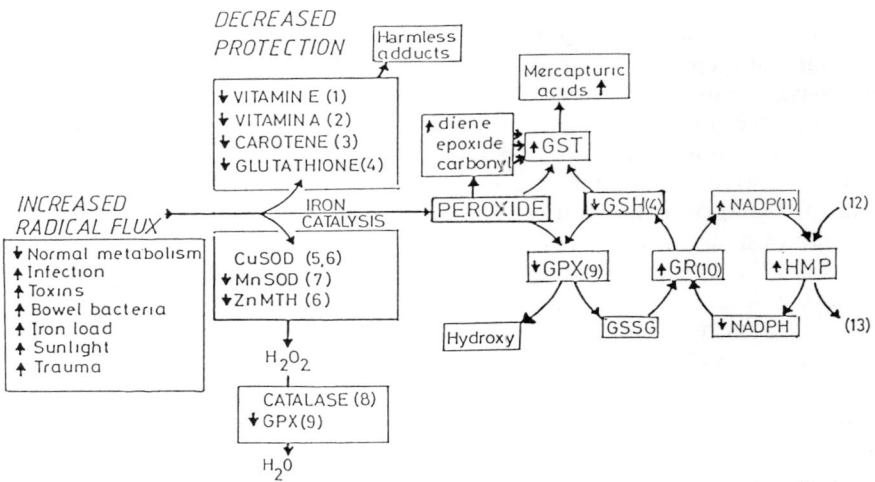

Fig. 1.   Diagram showing the mechanisms of radical production and sub-sequent metabolism. ↑ or ↓ beside the substrates, products, and enzymes shows whether they have been demonstrated to be increased or decreased in children with kwashiorkor. The essential nutrients involved are: (1) Vitamin E; (2) Vitamin A; (3) Carotene; (4) Sulphur amino acids; (5) Copper; (6) Zinc; (7) Manganese; (8) Iron; (9) Selenium; (10) Riboflavin; (11) Nicotinic Acid; (12) Magnesium and phosphorus; (13) Thiamin.

CuSOD: Cu—Zn superoxide dismutase; Mn—SOD: Manganese superoxide dismutase; Zn-MTH:Zn-metallothionein; GST: Glutathione S-transferase; GPX: Glutathione peroxidase; GSH: Glutathione (reduced); GSSG: Glutathione (oxi-dised); GR: Glutathione reductase; HMP: Hexose-monophosphate shunt (glucose-6-phosphate dehydrogenase, 6-phosphogluconic acid dehydrogenase); CAT: Catalase. (24)

2. MALNUTRITION

Severe malnutrition is clearly associated with poverty. It may be thought that the resulting conditions of marasmus and kwashiorkor are simply the result of starvation and lack of intake of the bulk nutrients. Except in famine situations, such as we have recently seen in sub-Saharan Africa, this is rarely the case. When one visits the home of a malnourished child, there is nearly always "food" available (46). The usual malnourished child has anorexia. Repeatedly, mothers give histo-ries of children refusing feed when offered, even breast milk. Most children, particularly those with kwashiorkor, when admitted to hospi-tal, have to be force-fed initially. Those who have not cared for these patients imagine that they are ravenous. This is not the case. Why should a child, with food available, refuse to eat it if he's "starving"? In famine situations, the patients are quite different; they are usually demonstrably hungry except when the condition is complicated.

What characterizes poverty is **not** a lack of food *per se*, so much as

Table 1

Nutrients that are Important in Dismutation and Protection for Free Radicals

Essential nutrients	Function
1. Tocopherol	the body's natural fat soluble antioxidant
2. Retinol	
3. Ascorbic acid	natural water-soluble antioxidant
4. Sulphur amino acids	glutathione, taurine, mercapturic acid formation
5. Copper	copper-zinc superoxide dismutase
6. Zinc	copper-zinc superoxide dismutase, membrane stabilisation metallothionein
7. Manganese	manganese superoxide dismutase (mitochondrial protection)
8. Iron	catalase
9. Selenium	glutathione peroxidase; other function in sulphydryl protection
10. Riboflavin	co-enzyme for glutathione reductase
11. Nicotinic acid	NADPH, required by glutathione reductase
12. Thiamine	transketolase (hexose monophosphate shunt)
13. Magnesium	required by hexose monophosphate shunt and glutathione synthesis
14. Phosphorus	required by hexose monophosphate shunt
Other antioxidants	
15. Carotene	
16. Uric acid	

lack of variety of foods. Rich people have a very varied diet. Even the sources and varieties of the 'same' foods are heterogeneous. They contain different dishes at most meals, each dish composed of a large variety of foods. Nutritionally, variety is indeed the spice of life. As a family becomes poorer, the luxury items are first dropped from the diet, and the more expensive "common" foods rarely consumed. Cheaper varieties from the least expensive sources are chosen. With increasing poverty, the diet becomes more and more restricted until, perhaps, only the staple food is being consumed. The variety of the diet can be assessed by food frequency analysis and Guttman Scaling (12). Children are the most vulnerable, not only because they are immunologically naive and have relatively high requirements for nutrients, but also because they are the most likely members of the family to be given the same food day in, day out. In Jamaica, whereas a parent is eating different foods, it is common for the child to be given maize-meal porridge with tea at every meal. **Monotony is the hallmark of poverty.**

It is difficult for the middle-class and those in the Western World to appreciate just how monotonous it is to sit down to a bowl of rice at **every** meal for months or years on end. Monotony is also the hallmark of poor nutrition. The quantitation of 'monotony' should form an integral part of nutritional assessment; however, we do not have adequate tools to do this. Guttman scaling, which is the best tool available, is hardly ever used in dietary assessment: it can be very useful (3). It is monotony that characterizes the diets of the isolated elderly and the disadvantaged in the developed world, as well as in the developing world; they also become malnourished, though not necessarily anthropometrically. Almost no food is, by itself, nutritionally adequate. Even human breast milk, if given alone for long enough, will lead to nutrient deficiency, in this case, of copper and iron. All other foods are individually deficient in some respect. As the dietary becomes more and more restricted, the nutrient intake becomes increasingly unbalanced until overt deficiencies occur; in some situations, of one nutrient, in others, of many. Because there are so many nutrients involved in the free radical protective repertoire, it is likely that it will be compromised in many situations, which lead to 'poor nutrition.'

Most nutrient deficiencies (as well as infections and liver disease) lead to anorexia. There is little point in continuing to eat the diet that led to the deficiency in the first place! If it is offered, it is refused. Hence, anorexia is the usual harbinger of severe malnutrition. After the patient refuses to eat, weight loss and deterioration are rapid. Indeed, malnutrition (used in the pure sense of the word) probably **precedes** loss of weight and the other anthropometric changes we often use to define the malnourished state.

In primary starvation, the situation is completely different: the patient effectively 'eats' his own tissues, a very good diet until the body stores are exhausted: that patient is indeed hungry. It is interesting that in anorexia nervosa, in which the diet is usually varied but very restricted in amount, the patients are alert, active, and usually without any obvious clinical or biochemical sign of a specific deficiency, except for profound weight loss. The biochemistry of these patients is, thus, totally different from those with malnutrition secondary to disease or malnutrition associated with poverty, except in the most extreme cases.

It is clear from these differences between anorexia nervosa or starvation and the other common forms of malnutrition that malnutrition is not simply the result of an insufficient energy intake. An important corollary of this is that "sickness," morbidity and mortality, may not bear a close relationship to weight loss or other anthropometric changes. Indeed, this is what is found in practice. Anthropometric changes, or the lack of such changes, are a poor reflection of nutritional health.

If the malnourished child, subsisting on a monotonous diet with relatively low levels of the nutrients involved in free radical protection, is the most obvious situation in which to search for free radical mediated disease in man, what is the evidence that damage actually occurs? In other words, what is the evidence that the radical production is greater than the capacity of the antioxidant defense system in this situation?

Of course, the essential nutrients in Table 1 are involved in many processes in the body apart from protection against free radicals. Indeed, many of these nutrients are involved in free radical production! Consequently, it is not immediately obvious what the resultant effects of a dietary deficiency will be. The situation is likely to be extremely complex, with the outcome dependent on the balance of effects. These are likely to differ between different organs and different radical producing stimulae, and depend on the relative vulnerability of each system to a lack of the nutrient under consideration. The relative vulnerability of the generating and protective components will, in turn, depend on the synthesis rate and breakdown rate of the component coordinating molecules (and hence the acuteness of the deficiency), on the other coincidental nutrient deficiencies and on the ability of the body to control whichever of the two effects will predominate. These factors will be different for each nutrient.

For example, nicotinic acid functions in the form of NAD(H) and NADP(H). NADPH, the reduced form, is required for the reduction of oxidized glutathione by glutathione reductase. It is thus crucial to the maintenance of the pool of GSH, and consequently, to the ability of gluthathione peroxidase and glutathione S-transferase to function. Without GSH, the body's sulphydryls become oxidized. On the other hand, NADPH is used to provide reducing equivalents for cytochrome-p_{450} and for white cell lipooxygenases, both of which are primary sources of radicals. Thus, in nicotinic acid deficiency, one may predict a reduction in radical production by cytochrome-p_{450} and the lipooxygenases, as well as a concomitant decrease in GSH, in many organs. Perhaps, this accounts for the fact that the skin is photosensitive and damaged in pellagra, whereas the intestinal mucosa is the other organ that sustains major damage. These two organs are exposed to continuing radical flux that is independent of the presence or absence of NAD(P): the low GSH is the important determinant in tissue damage. The liver is not a target in this particular deficiency. This may be because of the concomitant diminution in the cytochrome-p_{450} activity. Thus, it could be hypothesized that, despite a reduction in hepatic GSH, the rate of radical production is also reduced to a degree where hepatic damage does not occur, at least without a very severe deficiency. With this type of consideration, we may be able to explain why isolated deficiencies of the individual nutrients involved in radical protection have quite different clinical expressions in different 'target' tissues.

Similarly, iron deficiency may lead to a reduction in catalase. This is part of the protective system. However, it is a relatively minor and unimportant part, except in the peroxisome—a specific subcellular organelle involved, inter alia, with fat oxidation. In other subcellular sites, glutathione peroxidase performs the same reaction. The reduction in cytochrome-p_{450}, the lipooxygenases, and ferritin, all of which contain iron and are potential major sources of free radicals, is likely to outweigh the reduction in catalase activity except in peroxisomes. Thus, iron deficiency may well lead to a net reduction in oxidative stress, notwithstanding a reduction of catalase.

These two examples illustrate the complexity of nutrient deficiency with respect to free radical mechanisms, how nutrients may behave differently, and why effects may differ between and within organs and subcellular compartments.

Zinc, copper, iron, and selenium are all intimately involved in the immune and inflammatory systems. As immune stimulation of white cells is a major source of free radicals, the loss of protection that occurs with deficiency of these nutrients will be naturally buffered by the concomitant reduction in radical production by the immune system. Of course, a diminution in the immune response does not necessarily mean that there will be an overall reduction in radical production; only, that the source of radicals will be different. They will arise from cells damaged by the invading organisms and their toxins rather than from the host's white cells. Indeed, with a compromised immune system, which should kill the organisms when they are few and isolated, the invaders may become widespread and numerous. This will give such a stimulus to the defective immune system that endogenous radicals may be widely produced throughout the body from a maximally stimulated immune system, as well as from the products of directly damaged cells. None of the relationships is straight forward, and few can be predicted theoretically.

These arguments are, of necessity, speculative. There are very few data. Few investigators have examined either radical production, antiradical defense, or the various types of nutritional deficiency and interaction from these points of view, in experimental animals, let alone, in human populations. It is likely, however, that these types of consideration underlie the different presentations of diseases, such as measles, in Europe and Africa. In Europe, measles is a relatively benign self-limiting exanthem. In Africa, it has a very high mortality rate, but, there is very little in the way of a rash or local reaction to the measles virus. The differences in presentation and severity are almost certainly nutritionally-mediated, and probably involve radical production and protection.

In this way, severely malnourished populations may constitute a group that is susceptible to the damaging effects of excess free radical

Table 2
Observations Made in Children with Kwashiorkor

1. Fatty liver, hair and skin changes, immunoparesis and mental changes are associated with oedema.

2. Whereas wasting of tissues may be present, the patient is rarely severely stunted.

3. Patients are usually infected and have an overgrowth of bacteria in their small intestine.

4. Skin lesions are particularly florid in dark skinned patients in the tropics. They have histological features similar to that of pellagra and zinc deficiency.

5. There is resolution of clinical signs on a diet containing a lower concentration of protein than the antecedent diet, with no change in plasma albumin.

6. There is an epidemiological association with specific foods: cassava, yam, plaintain, maize, rice, cruciferae.

7. Kwashiorkor is precipitated by many different infections, particularly measles, tuberculosis, malaria, diarrhoea.

8. There are low plasma levels of transport proteins (albumin, transferrin, ceruloplasmin, retinol binding protein), vitamin E, vitamin A, carotene, zinc, copper and selenium.

9. Red cell glutathione and glutathione peroxidase are low.

10. Red cell hexose monophosphate shunt, Na^+K^+ ATPase, glutathione reductase and glutathione S-transferase are high.

11. There are high levels of circulating ferritin. Hepatic iron is high, whilst other trace elements are low.

12. There is no animal model of kwashiorkor.

production. In underdeveloped countries, a large proportion of children suffers from malnutrition. This usually leads to stunting and marasmus, and in a small percentage of children, to kwashiorkor (*21–23*).

Stunting and marasmus are probably the end results of a qualitatively inadequate diet, together with infection (*22*). However, the underlying conditions that precipitate the clinical picture of kwashiorkor are largely unknown. Kwashiorkor is characterized by edema, severe fatty liver, skin dyspigmentation and breakdown, hair color changes, and very low levels of circulating hepatic export proteins (*21,24,69*). Other clinical and biochemical features of this disease are given in Table 2.

We have put forward the argument that kwashiorkor results from an imbalance between the production of free radicals and their safe disposal (*21,24*). The proposed unifying hypothesis, which would account for all the features described, is that the various noxae to which these children are ubiquitously exposed, produce an oxidative stress that leads to excess free radical, peroxide, and carbonyl generation. These toxic products cause damage because the mechanisms for their safe disposal are generally compromised (24). The impaired state of the body's antioxidant defense is directly caused by the inadequate diets that these children consume (21,61). What evidence is there to support the existence of oxidative stress in these children?

Oxidative stress represents a common end point, and is particularly difficult to quantitate: if the stress is sufficiently severe, normal protective mechanisms will be overwhelmed. Alternatively, if the diet is sufficiently poor (as in the case of the malnourished child), even normal rates of radical formation may be damaging. Moreover, if the stress is increased and protection decreased, there is likely to be damage. The question whether antioxidant systems are compromised will be examined in the following section.

2.1. Lipid Soluble Antioxidants

Tissue levels of several of the antioxidant nutrients, listed in Table 1, have been measured in children with kwashiorkor (24). Thus, plasma levels of vitamin E (11,24,47) and carotene (48,65) are markedly reduced, both absolutely and relative to lipid concentrations, in children with kwashiorkor. The magnitude of the deficiencies of these chain breaking antioxidants is related to prognosis, and increases across the clinical spectrum, going from marasmus to kwashiorkor (24,48). Antioxidant protection in lipophilic domains is, therefore, compromised in children with kwashiorkor.

Oxidized vitamin E is reconverted to the active form at the expense of ascorbic acid (39,56). Given the low levels of vitamin E in malnourished children, plasma levels of ascorbic acid may also be low in these children, as the ascorbate will be consumed to regenerate vitamin E. In the few studies that have been carried out in malnourished children, plasma ascorbate levels were marginally reduced (43). Ascorbic acid, on the other hand, can provide reducing equivalents for iron redox cycling, and thus, in the presence of excess iron, may be a prooxidant!

2.2. Sulphur Amino Acids

Glutathione plays a crucial role in maintaining the sulphydryl status of cells. It is an integral part of several metabolic pathways, particularly those involved in detoxification (36,38,44). As such, glutathione is an essential component of the cellular antioxidant defense: it is frequently used as an index of prooxidant stress (38,64). Figure 2 shows whole blood glutathione concentrations of malnourished children. Those with marasmus have normal concentrations of glutathione: there is virtually no overlap with those of children suffering from kwashiorkor. This lack of overlap is unlike all other biochemical indices previously measured. Here, we have used three control groups: healthy children of similar ages to the children with kwashiorkor, marasmic children as nutritional controls, and children with nephrotic syndrome as edematous controls. In

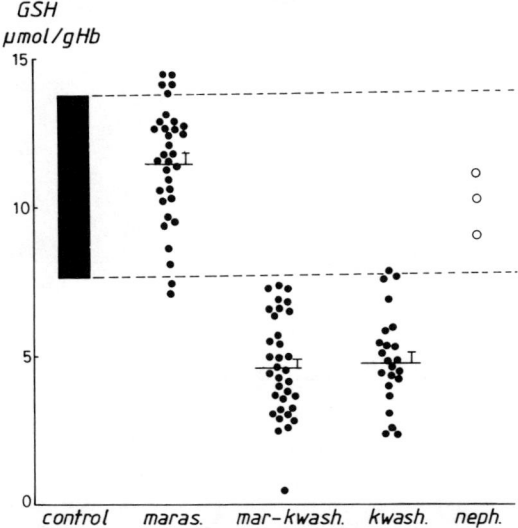

Fig. 2.   Whole blood glutathione (GSH & GSSG) levels separated according to diagnosis on admission. Maras = Marasmic; Mar-kwash = Marasmic
kwashiorkor; Kwash = Kwashiorkor; Neph = Nephrotic syndrome. Means and
SEM are shown for each group.

each control group, the glutathione levels are normal. Thus, glutathione
is specifically decreased in children with kwashiorkor, and bears a strong
inverse relationship ($r = 0.7$) with the degree of edema on admission.
Further, it is possible to diagnose kwashiorkor independently, using a
glutathione range of < 6.8 mol/gHb, with a sensitivity, specificity, and
positive accuracy of greater than 90%. On recovery from kwashiorkor
glutathione levels become normal.

It is clear that differences in the dietary intake of sulphur amino acids
cannot account for the difference in glutathione levels: children with
marasmus and kwashiorkor seem to have the same antecedent diet.
Although these diets are probably low in sulphur amino acids, intakes
are sufficient to maintain normal levels of glutathione in the marasmic
children. Alterations in the metabolic pathways that involve glutathione
suggest that the diminution of glutathione levels, in children with
kwashiorkor, is likely to be the result of increased consumption (21,24).

2.3. Metabolic Pathways Involving Glutathione

The seleno-enzyme, glutathione peroxidase (GPx), is actively involved in the removal of hydroperoxides (34). This enzyme is dependent
on both dietary selenium and glutathione. Children with kwashiorkor
have a particularly poor selenium status (9,20,40,53). Furthermore, analysis of red cell density fractions for glutathione peroxidase shows that the
light (young) cells have **less** enzyme than the heavier cells. This suggests

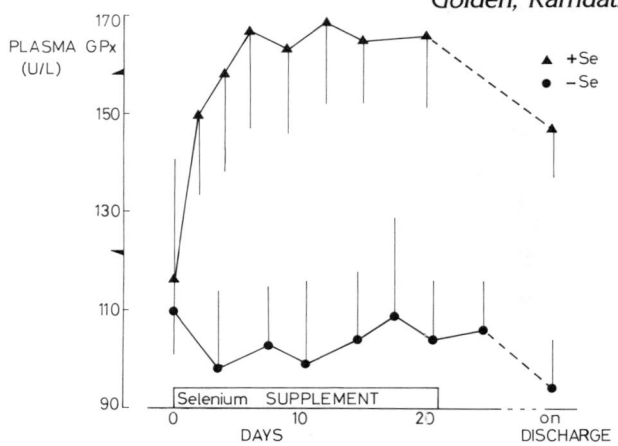

Fig. 3.   Plasma glutathione peroxidase (GPx) activity during recovery from malnutrition (means ±SEM). Closed circles show GPx activity on a low selenium cow's milk based diet. Closed triangles show GPx activity when children were given sodium selenate (20 μg Se/kg/d) along with the basal diet. (53)

that the immediate antecedent diet was selenium deficient. (Golden and Ramdath, unpublished). Figure 3 shows that when given supplements of sodium selenate, these children show a rapid increase in plasma GPx activity. A similar response in plasma selenium was also measured (53).

Given the reduced activity of erythrocyte GPx, it is reasonable to expect a compensatory change in the other components of the antioxidant repertoire. One of the alternative mechanisms for detoxification is the glutathione S-transferase family of isoenzymes (4,36,44,45). Thus, erythrocyte glutathione S-transferase (GST) activity is elevated in over 50% of malnourished children (58). This induced activity of GST is probably the result of substrate induction of the enzyme, combined with selenium deficiency (49).

The end products of lipid peroxidation include several complex carbonyls (18). Detoxification of these compounds by GST involves the formation of a glutathione adduct that is sequentially degraded and excreted in the urine as a mercapturic acid (28,55). Urinary mercapturic acids (UMCA), therefore, provide a semiquantitative measure of the body's burden of GST substrates. Figure 4 shows that, compared to normal, UMCA outputs are significantly increased in both kwashiorkor and marasmus. This implies that the toxic stress to which children are exposed, in vivo, is abnormally high. Moreover, it demonstrates that glutathione is being consumed at an increased rate.

Why are the UMCA outputs increased in children with marasmus as well as in those with kwashiorkor? These children presumably experience the same noxious stresses to which children with kwashiorkor are exposed. Given that glutathione levels are normal in marasmic children, they are, at the time of the stress, better equipped with antioxidants: it is

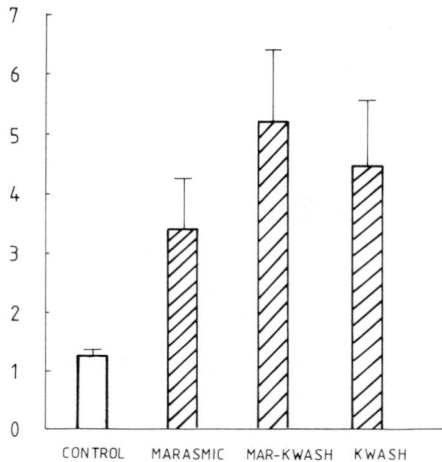

Fig. 4.   Urinary total mercapturic acid (UMCA) outputs in severely malnourished children (on admission), and in a group of healthy children (57).
Means and SEM.

not unusual for marasmic children to subsequently develop kwashiorkor
as well—so-called marasmic-kwashiorkor. Children with kwashiorkor
may be unable to maintain cellular glutathione in the functional, reduced
form (GSH). Figure 5 shows the proportion of erythrocyte $NADP_{tot}$
$[NADP^+ + NADPH]$ that exists in the oxidized form $[\%NADP^+/NADP_{tot}]$;
as measured longitudinally in children recovering from marasmus and
kwashiorkor. Those with marasmus have normal $\%NADP^+/NADP_{tot}$;
this remains constant throughout recovery. Children with kwashiorkor
have a larger proportion of nucleotide in the oxidized form (58%). We
interpret these data to show: (a) that the cellular NADPH-redox is altered
and (b) the cellular environment of children with kwashiorkor is oxidized. The oxidized cells require approx 9 d to return to normal. Under
these redox conditions, sulphydryl groups, including those of GSH, will
become oxidized. As intracellular levels of oxidized glutathione (GSSG)
increase, the GSSG is actively exported from the cell (66): this immediately leads to a reduction of glutathione concentration unless it is
resynthesized at a rate sufficient to compensate for the loss. GSSG can be
converted to GSH by glutathione reductase (GR) and reduced NADPH
(from the hexose monophosphate shunt).

Table 3 shows that the enzymes involved in maintaining normal
levels of erythrocyte GSH [glucose-6-phosphate dehydrogenase: G6PD;
6-phosphogluconic acid dehydrogenase: 6PGD; and erythrocyte GR] are
all operating at an elevated level in children with malnutrition—the
wheels are turning faster because they are being driven by substrate.

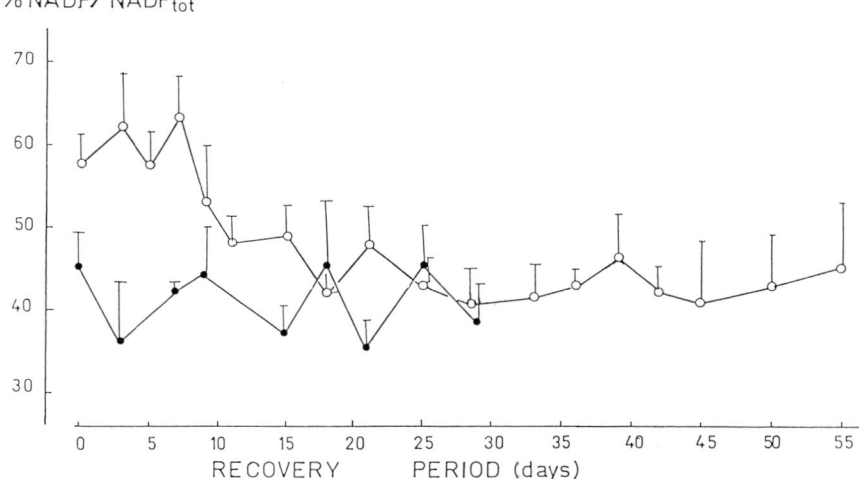

% NADP⁺/NADP_tot

Fig. 5.   The proportion of total nucleotide in the oxidized form (%NADP⁺/NADP_tot) in erythrocytes from children recovering from edematous malnutrition (open circles) or marasmus (closed circles). Means and SEM are shown. (D. D. Ramdath & M. H. N. Golden, unpublished).

Table 3
Activities of Erythrocyte Enzymes Involved
in the Maintenance of Normal Glutathione Status

	Enzyme Activity in Units/g Hb at 25 °C		
	G6PD	6PGD	EGR*
Controls	5.63±0.64 (18)	3.01±0.22 (18)	6.63±0.23 (22)
Marasmus	6.86±0.33 (95)	6.04±0.22 (96)	8.95±0.46 (23)
Kwashiorkor	7.00±0.44 (65)	6.45±0.40 (64)	7.66±0.53 (17)

Mean and SEM are shown. Values in parenthesis show the number of subjects. G6PD and 6PGD activity assayed by the method of Bishop (8). *Data from Golden et al. (26)

Collectively, these data suggest that malnourished children are able to adapt to the chronic toxic stress to which they are exposed, given time and a supply of the essential nutrients. This is accomplished by modulation of enzyme activity and compensation in various facets of the antioxidant system. It should be emphasized that the red cell enzyme changes are all chronic, adaptative changes. Unlike the changes in enzyme activities, cytoplasmic substrates, such as GSH and NADPH, reflect very recent changes in the cellular redox status. The fact that these are only altered in children with kwashiorkor suggest that this disease may be precipitated by a sudden, but intense prooxidant stress. It is, therefore, not surprising that the incidence of kwashiorkor increases following epidemics of measles (52). Here, measles is a single example of numerous situations that could present as prooxidant stress. It is important in this context that the history of kwashiorkor is nearly always short. Textbook descriptions of the disease neglect this point: they simply describe the grossness of the clinical picture, and the assumption is made that it must have taken a considerable time to reach that stage. This is not so. The skin lesions and edema develop over a few days. We have occasionally observed the development of the full syndrome in a child with an admission diagnosis of marasmus. It develops in 24 to 48 h. This agrees with the history given by the parents of nearly all the patients that we admit. It also provides a perfectly adequate explanation why the enzyme levels are not different in marasmic and kwashiorkor children, whereas the GSH and NADPH levels are markedly different. Given increased stress, many of the marasmic children are at risk of developing marasmic-kwashiorkor.

3. SOURCES OF PROOXIDANT STRESS

3.1. Infections

Children with kwashiorkor are frequently infected with one or more microorganisms (13,51). The normal response to invading organisms is an enhanced production of several species of free radical by the stimulated neutrophils (5,6). In the healthy individual, these bactericidal species are quickly dissipated by an active antioxidant system. In the malnourished child, the reaction is much more damaging.

3.2. Small Bowel Overgrowth

Bacterial overgrowth of the small bowel is ubiquitous in children with kwashiorkor (32) and may be a major source of endotoxin and bacterial metabolites (17). Fatty liver, the hallmark of hepatic free radical damage, is produced by small bowel overgrowth, both clinically and experimentally (15,54).

3.3. Dietary Toxins

Children living in areas where kwashiorkor is prevalent ingest foods that are heavily contaminated with bacteria (7,33,62). Similarly, Hendrickse et al. (31) have reported high levels of aflatoxin and its metabolites in plasma and liver samples obtained from African children. The level of prooxidative insult, to which children with kwashiorkor are exposed, is overwhelming. When this is related to the impairment

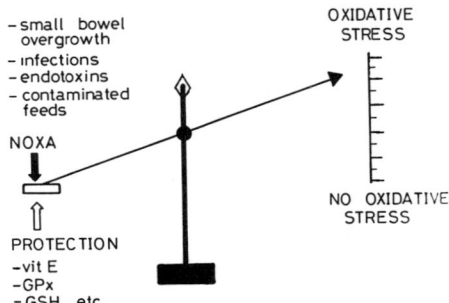

Fig. 6.   Diagram showing the proposed imbalance between antioxidant protection and prooxidant stress, in favor of the latter, and the resultant oxidative stress to which children with kwashiorkor are exposed.

of antioxidant function, it becomes clear that kwashiorkor is associated with a severe and continuous oxidative stress. This concept is outlined in Fig. 6.

4. IRON AS A CATALYST
OF FREE RADICAL REACTION

Compounding the problem of prooxidant stress in children with kwashiorkor, is the possibility that iron takes part in free radical propagation, in vivo. Plasma levels of iron are normal in kwashiorkor; however there is usually a significant reduction in plasma transferrin. Thus, the total iron binding capacity of plasma is abnormally decreased in these children (59,60). Table 4 compares plasma levels of iron and iron binding capacity, as reported by various groups working in the developing world. Clearly, if the plasma iron levels are normal, and transferrin levels low, then transferrin saturation will be high. This is precisely what is found: transferrin saturation is high in kwashiorkor, and is related to mortality (42,59,60).

4.1. Iron Stores

Increased levels of plasma ferritin have been reported in children from several geographical locations where kwashiorkor occurs (10,25,27,59). In particular, Golden et al. (25) have shown a direct relationship between plasma ferritin levels and mortality. We have argued that the high ferritin values found in children with kwashiorkor reflect high body stores of iron, and are not simply a response to infection and inflammation (25,59). The proportion of plasma ferritin glycosylated is normal, that is, 60–80% (35). On recovery from kwashiorkor, when infections are absent, plasma ferritin levels remain elevated (25,59). Further, direct measurements of trace elements in children who have died

Table 4
Plasma Iron and Iron Binding Capacity
in Children with Oedematous Malnutrition

	Plasma iron μg/dL*	Plasma iron binding capacity (μg/dL)[#]
Adams and Scrapp (1)	54 (7–102)	95 ± 43
Sandstead et al. (63)	49 (12–91)	91 ± 52
Lynch et al. (41)	55 (17–93)	154 ± 97
Ramdath and Golden (59)	76 (15–116)	124 ± 123
Normal	75 (31–119)	355 ± 58

*—Mean and range. #Mean and standard deviation (59).

from kwashiorkor show that only iron and sodium are present at higher than normal concentrations (50,59,68).

4.2. Is Mobilizable Iron Present in Children with Kwashiorkor?

High transferrin saturation, high plasma levels of ferritin, the presence of stainable iron in the bone marrow, and excess hepatic iron in children who have died, are all indicative of iron overload in kwashiorkor.

However, we needed independent confirmation of high iron stores, using a method that is not subject to the errors of ferritin or the bias of postmortem material. The iron chelating drug, desferrioxamine (DFO), is presently used for treatment of iron overload. Its administration to patients with various forms of iron overload results in the urinary excretion of iron (14). Under normal circumstances, this is negligible (16). Urinary iron excretion following DFO administration can, therefore, be regarded as an independent measure of the amount of iron in the body that can be easily mobilized. In other words, it is a measure of both the iron burden and the easily mobilized iron, potentially available for redox cycling.

Table 5 shows 24 h urinary iron output following 500 mg intramuscular DFO, along with the corresponding plasma ferritin values. In the groups diagnosed as having kwashiorkor or marasmic–kwashiorkor, the mean urinary iron output was much higher than that for children with marasmus. However, the question of normal values arises, as these have not been determined for this pediatric age range.

Regression of the DFO urinary iron against plasma ferritin gives a highly significant relationship; from this equation, the urinary excretion that corresponds to the normal pediatric range of ferritin is 30 to 400 μg/24 h. In the adult, with his much larger iron pool, the excretion of 1000 μg after DFO is regarded as diagnostic of iron overload. These

Table 5
Urinary Iron Excretion Following Desferrioxamine

TYPE OF MALNUTRITION	URINARY IRON ug/24hr	PLASMA FERRITIN ug/l
MARASMUS n=8	367 ± 90 (103 – 836)	82 ± 40 (12 – 355)
MARASMIC KWASHIORKOR n=11	708 ± 224 (70 – 2465)	300 ± 128 (13 – 1250)
KWASHIORKOR n=10	987 ± 219 (582 – 2862)	329 ± 222 (12 – 2320)

Means and SEM are reported; range is given in parenthesis. Plasma ferritin was assayed using an enzyme immunoassay as per "Spectro Ferritin" method (Ramco Laboratories, Texas). Desferrioxamine (500 mg) was given intramuscularly. (Data from Ramdath and Golden (59)

considerations confirm that children with edematous malnutrition are, as a group, suffering from iron overload. Obviously, they should not be given therapeutic doses of iron during their acute illness. Indeed, one may argue that these children should be treated initially with DFO until infections are treated, the iron binding proteins are resynthesized, and they have an increased demand for iron during 'catch-up' growth.

Using the urinary iron excretion following DFO as a measure of "iron stress" and the close relationship between plasma ferritin and urinary iron (35), the following equation was derived to predict the "iron stress" of malnourished children from plasma ferritin and plasma transferrin:

$$\text{IRON STRESS} = \exp(6.1557 + 0.0087\,X_a - 0.0041X_b)$$

where:

iron stress = theoretical value for the urinary iron excretion;

X_a = plasma ferritin; X_b = plasma transferrin.

4.3. Effects of Iron Overload

Elevated iron stores increase the possible occurrence of free iron in vivo (37). This is associated with lipid peroxidation, the production of cytotoxic compounds, and subsequent membrane damage (18,19,29, 30,67). It is precisely this sort of damage, in which the function of the protein synthetic machinery, the packaging and export of lipoproteins, and the transfer of acyl groups into the mitochondrion are globally impaired, that one would expect to find grossly diminished hepatic export protein levels and fatty accumulation in the liver: these are the hallmarks of kwashiorkor.

5. CONCLUSION

People who live in poverty have stress as part of their daily life. This is not only psychological; it is also dietary, physical, infectious, and biological. It is indeed difficult for those who only observe, to comprehend what this burden is like. The insults compound one another. Much of the basic biochemistry of the body surrounds the reactions catalyzed by components that generate free radicals: free radicals are part of life. How these highly reactive species are affected and involved in the diseases of poverty is only beginning to be unraveled. We have sufficient information from the malnourished child to realize that that involvement is likely to be central to the expression, or nonexpression, of the results of the insults of poverty.

ACKNOWLEDGMENTS

The work reported was fully supported by the Wellcome Trust. We thank the nurses and technical staff of the Trace Element Research Group for their careful and continued contribution to this work.

REFERENCES

1. E. B. Adams, and J. N. Scragg. Iron in the anaemia of kwashiorkor. *Br. J Haematol*. **2**, 676–681 (1965).
2. G. Arroyave, D. Wilson, J. Mendez, M. Behar, and N. S. Scrimshaw. Serum and liver vitamin A and lipids in children with severe protein malnutrition *Am. J. Clin. Nutr*. **9**, 180–185 (1961)
3. P. Arroyo, S. E. Q. De Arroyo, S. E. Perez Gil, and A. Chavez. Correlation between family and infant food habits by scalogram analysis. *Ecology of Food and Nutrition* **1**, 127–130 (1972).
4. Y. C. Awasthi, and S. V. Singh. Purification and characterization of a new form of glutathione *S*-transferase from human erythrocytes. *Biochem. Biophys. Res. Commun*. **125(3)**, 1053–1060 (1984).
5. B. M. Babior. Oxygen-dependent microbial killing by phagocytes. *N. Engl. J. Med*. **298(12)**, 659–668 (1978a).
6. B. M. Babior. Oxygen-dependent microbial killing by phagocytes. *N. Engl. J. Med*. **298(13)**, 721–725 (1978b).
7. R. A. E. Barrell, and M. G. M. Rowland. Infant food as a potential source of diarrheal illness in rural West Africa. *Trans. Roy. Soc. Trop. Med. Hyg*. **73(1)**, 85–89 (1979).
8. C. Bishop. Assay of glucose-6-phosphate dehydrogenase (E.C. 1.1.1.49) and 6-phosphogluconate dehydrogenase (E.C. 1.1.1.43) in red cells. *J. Lab. Clin. Med*. **63(1)**, 149–155 (1966).
9. R. F. Burk, W. N. Pearson, R. P. Wood, and F. Viteri. Blood selenium levels and in vitro red blood cell uptake of ^{75}Se in kwashiorkor. *Am. J. Clin. Nutr*. **20**, 723–733 (1967).

10. B. Caballero, N. S. Solomons, R. Batres, and B. Torun. Homeostatic mechanisms in the utilization of exogenous iron in children recovering from severe malnutrition. *J. Ped. Gastroenterol. Nutr.* **4**, 97–102 (1985).
11. L. Charley, J. Foreman, D. Ramdath, F. Bennett, B. Golden, and M. Golden. Vitamin E in malnutrition. *West Indian Med. J.* **34(suppl)**, 62–63 (1985).
12. J. Chassy, A. G. Van Veen, and F. W. Young. The application of social science research methods to the study of food consumption in an industrializing area. *Amer. J. Clin. Nutr.* **20**, 56–64 (1967).
13. C. Christie, G. T. Heikens, and D. E. MacFarlane. Infections in malnourished children. *West Indian Med. J.* **34(suppl)**, 47 (1985).
14. A. Cohen. Management of iron overload in the pediatric patient. *Ped. Hematol.* **1(3)**, 521–543 (1987).
15. M. S. Cooperstock, R. P. Tucker, and C. G. Baublis. *Lancet* **i**, 1272–1274 (1975).
16. J. H. Dagg, J. A. Smith, and A. Goldberg. Urinary excretion of iron. *Clin. Sci.* **30**, 495–503 (1966).
17. B. S. Drasar, and M. J. Hill. *Human Intestinal Flora*. Academic New York, (1974).
18. H. Esterbauer. Lipid peroxidation products: formation, chemical properties, and biological activities. In *Free Radicals in Liver Injury*, (G. Poli et al., eds.) IRL Press Ltd., Oxford, 1985, pp. 29–47.
19. H. Esterbauer, K. H. Cheeseman, M. V. Dianzani, and T. F. Slater. Separation and characterization of the aldehydic products of lipid peroxidation stimulated by ADP-Fe^{2+} in rat liver microsomes. *Biochem. J.* **208**, 129–140 (1982).
20. P. Fondu, C. Hariga-Miller, N. Mozes, J. Neve, A. van Steirteghem, and L. M. Mandelbaum. Protein-energy malnutrition and anemia in Kivu. *Am. J. Clin. Nutr.* **31**, 46–56 (1978).
21. M. H. N. Golden. The consequences of protein deficiency in man and its relationship to the features of kwashiorkor. In *Nutritional Adaptation in Man*, K. Blaxter and J. C. Waterlow, eds., Applied Science Publishers, London, 1985, pp. 169–187.
22. M. H. N. Golden. The effects of malnutrition in the metabolism of children. *Trans. Roy. Soc. Trop. Med. Hyg.* **82**, 3–6 (1988).
23. M. H. N. Golden. The role of individual nutrient deficiencies in growth retardation of children as exemplified by zinc and protein. In *Linear Growth Retardation in Less Developed Countries*, J. C. Waterlow, ed., Nestle Nutrition Workshop Series #14. Vevey/Raven, New York, 1989, pp. 143–163.
24. M. H. N. Golden, and D. Ramdath. Free radicals in the pathogenesis of kwashiorkor. *Proc. Nutr. Soc.* **46**, 53–68 (1987).
25. M. H. N. Golden, B. E. Golden, and F. I. Bennett. High ferritin values in Jamaican malnourished children. In *Trace Element Metabolism in Man and Animals* #5, C. F. Mills, ed., Commonwealth Agricultural Bureaux, Slough, 1985, pp. 775–779.
26. B. Golden, D. Ramdath, J. Appleby, and M. Golden. Erythrocyte glutathione reductase activity and riboflavin status of malnourished children. *West Indian Med. J.* **35**, 44–50 (1987).
27. A. T. Guiro, M. G. Sall, O. Kane, A. M. Ndiaye, D. Diarra, and M. T. A. Sy. Protein-calorie malnutrition in Senegalese children. Effects of rehabilitation with pearl millett weaning food. *Nutr. Rep. Int.* **36(5)**, 1071–1079 (1987).
28. W. H. Habig, M. J. Pabst, and W. B. Jakoby. Glutathione transferases. The first enzymatic step in mercapturic acid formation. *J. Biol. Chem.* **249(22)**, 7130–7139 (1974).

29. B. Halliwell, and J. M. C. Gutteridge. Oxygen toxicity, oxygen radicals, transition metals, and disease. *Biochem. J.* **219**, 1–14 (1984).
30. B. Halliwell, and J. M. C. Gutteridge. The importance of free radicals and catalytic metal ions in human diseases. *Mol. Aspects Med.* **8**, 89–193 (1985).
31. R. G. Hendrickse, J. B. S. Coulter, S. M. Lamplugh, S. B. J. McFarlane, T. E. Williams, and M. I. A. Omer. Aflatoxins and kwashiorkor: a study in Sudanese children. *Br. Med. J.* **285**, 843–846 (1982).
32. B. Heyworth, and J. Brown. Jejunal microflora in malnourished Gambian children. *Arch. Dis. Child.* **50**, 27–33 (1975).
33. J. Hibbert, and M. H. N. Golden. What is the weanling's dilema? Dietary and fecal bacterial ingestion of normal children in Jamaica. *J. Trop. Paed.* **27**, 255–258 (1981).
34. W. G. Hoekstra. Biochemical role of selenium. In *Trace Element Metabolism in Animals #4*, W. G. Hoekstra, J. W. Suttie, H. E. Ganther, and W. Mertz, eds., University Park Press (Boston Standard), Baltimore, 1974, pp. 61–77.
35. M. A. Hudson-Thomas, D. Ramdath, and M. H. N. Golden. Iron in malnutrition. *West Indian Med. J.* **37(suppl)**, 26 (1988).
36. W. B. Jacoby. The glutathione S-transferases: a group of multifunctional detoxification proteins. In *Adv. Enzymol.* **36**, 383–414 (1978).
37. A. Jacobs. Low molecular weight intracellular iron transport compounds. *Blood* **50(3)**, 433–439 (1977).
38. N. S. Kosower, and E. M. Kosower. The glutathione status of cells. *Int'l. Rev. Cytol.* **54**, 109 (1978).
39. P. Lambelet, F. Saucy, and J. Loliger. Chemical evidence for interactions between vitamins E and C. *Experientia* **41(11)**, 1384–1388 (1985).
40. R. J. Levine, and R. E. Olson. Blood selenium in Thai children with protein-calorie malnutrition. *Proc. Soc. Expt'l Biol. Med.* **134**, 1030–1034 (1970).
41. S. R. Lynch, D. Becker, H. Seftel, T. H. Bothwell, K. Stevens, and J. Metz. Iron absorption in kwashiorkor. *Am. J. Clin. Nutr.* **23**, 792–797 (1970).
42. H. MacFarlane, S. Reddy, K. J. Adcock, H. Adeshina, A. R. Cooke, and J. Akene. Immunity, transferrin, and survival in kwashiorkor. *Br. Med. J.* **2**, 268–270 (1970).
43. A. S. Majaj. Vitamin E-responsive macrocytic anaemia in protein-calorie malnutrition. Measurements of vitamin E, folic acid, vitamin C, vitamin B12, and iron. *Amer. J. Clin. Nutr.* **18**, 362–368 (1966).
44. B. Mannervik. The isoenzymes of glutathione transferase. *Adv. Enzymol.* **57**, 357–417 (1984).
45. C. J. Marcus, W. H. Habig, and W. B. Jacoby. Glutathione transferase from human erythrocytes. *Arch. Biochem. Biophys.* **188(2)**, 287–293 (1978).
46. L. J. Mata, R. A. Kromal, J. J. Urrutia, and B. Garcia. Effect of infection on food intake: perspectives as viewed from the village. *Am. J. Clin. Nutr.* **30**, 1215–1227 (1977).
47. P. M. Mathias. Vitamin E status of children recovering from severe malnutrition. *Proc. Nutr. Soc.* **41**, 143A (1983).
48. D. S. McLaren. Vitamin deficiencies complicating the severe forms of protein-calorie malnutrition, with special reference to vitamin A. In *Calorie Deficiencies and Protein Deficiencies*, R. A. McCance and E. M. Widdowson, eds., J & A. Churchill, London, 1968, pp. 191–199.
49. A. Melhert, and A. T. Diplock. The glutathione S-transferases in selenium and vitamin E deficiency. *Biochem. J.* **227**, 823–831 (1985).
50. J. Miles, M. H. N. Golden, D. Ramdath, and B. Golden. Hepatic trace elements in kwashiorkor. In *Trace Element Metabolism in Man and Animals*.

#6, L. S. Hurley, C. L. Keen, B. Lonnerdal, and R. B. Rucker, eds., Plenum, New York, 1987, pp. 497–498.

51. C. O. Morehead, M. Morehead, D. M. Allen, and R. E. Olson. Bacterial infections in malnourished children. *J. Trop. Paed.* **20**, 141–147 (1974).

52. D. Morley. The severe measles of West Africa. *Proc. R. Soc. Med.* **57**, 846–849 (1964).

53. C. Murphy, B. Golden, D. Ramdath, and M. H. N. Golden. Selenium status during recovery from malnutrition effects of selenium supplementation. In *Trace Element Metabolism in Man and Animals # 6*, L. S. Hurley, C. L. Keen, B. Lonnerdal, and R. B. Rucker, eds., Plenum, New York, 1987, pp. 11–12.

54. J. P. Nolan. Spontaneous endotoxinemia. *J. Ped. Gastroentol. Nutr.* **4**, 7–8 (1985).

55. K. Okajima, M. Inoue, K. Itom, S. Horiuchi, and Y. Morino. Interorgan cooperation in enzymatic processing and membrane transport of glutathione S-conjugates. In *Glutathione, Transport and Storage in Mammals*, Y. Sakamoto, ed., VNU Science Press BV, Netherlands, 1983, pp. 129–144.

56. J. E. Packer, T. F. Slater, and R. L. Willson. Direct observation of a free radical interaction between vitamin E and vitamin C. *Nature* **278**, 737–738 (1979).

57. D. D. Ramdath, and M. H. N. Golden. Urinary mercapturic acid outputs of severely malnourished children. *Proc. Nutr. Soc.* **47**, 7A (1988).

58. D. D. Ramdath, and M. H. N. Golden. Elevated glutathione S-transferase activity in erythrocytes from malnourished children. In *Medical, Biochemical, and Chemical Aspects of Free Radicals*, O. Hayaishi, E. Niki, M. Kondo, and T. Yoshikama, eds., Elsevier, Amsterdam, 1989a, pp. 567–570.

59. D. D. Ramdath, and M. H. N. Golden. Nonhematological aspects of iron nutrition. In *Nutrition Research Reviews. 2*, R. H. Smith, J. W. T. Dickerson, A. G. Low, and D. J. Milward, eds., Cambridge University Press, Cambridge, 1989b, pp. 25–49.

60. D. D. Ramdath, M. H. N. Golden, B. E. Golden, and R. Howell. Plasma iron, transferrin and transferrin saturation of severely malnourished Jamaican children. *West Indian Med. J.* **39(Suppl 1)** (1990), 42.

61. K. Rhodes. Two types of liver disease in Jamaican children III. *West Indian Med. J.* **31**, 157–178 (1957).

62. M. G. M. Rowland, R. A. E. Barrell, and R. G. Whitehead. Bacterial contamination in traditional Gambian weaning foods. *Lancet* **i**, 136–138 (1978).

63. H. H. Sandstead, M. K. Gabr, S. Azzam, A. S. Shuky, R. J. Weiler, O. M. El Din, N. Mokhtar, A. S. Prasad, A. El Hifney, and W. J. Darby. Kwashiorkor in Egypt II. Hematological aspects (the occurrence of a macrocytic anemia associated with low serum vitamin E and a wide range of serum vitamin B_{12} levels). *Am. J. Clin. Nutr.* **17**, 27–35 (1965).

64. H. Sies. Hydroperoxides and thiol oxidants in the study of oxidative stress in intact cells and organs. In *Oxidative Stress*, H. Sies, ed., Academic, London, 1985, pp. 73–90.

65. F. R. Smith, D. S. Goodman, M. S. Zaklama, M. K. Gagr, S. El Maraghy, and V. N. Patwardhan. Serum vitamin A, retinol-binding protein, and prealbumin concentration in protein-calorie malnutrition. I. *Am. J. Clin. Nutr.* **26**, 973–987 (1973).

66. S. K. Srivastava, and E. Beutler. The transport of oxidized glutathione from human erythrocytes. *J. Biol. Chem.* **244**, 9–16 (1969).

67. T. F. Slater, K. H. Cheeseman, M. J. Davies, K. Proudfoot, and W. Xin. Free radical mechanisms in relation to tissue injury. *Proc. Nutr. Soc.* **46,** 1–12 (1987).
68. J. C. Waterlow. Fatty liver disease in infants in the British West Indies. Med. Res. Council. Special Report Series No. 263. HMSO, London (1948).
69. C. D Williams. Kwashiorkor. *Lancet* **ii,** 1151–1152 (1935).

Conclusions

IVOR E. DREOSTI

During the last decade, free radical chemistry has entered the biological arena. No longer are free radicals considered by biochemists to be of concern only to radiation physicists and to industrial chemists. Now, it is known that free radical reactions occur widely in living tissue as a result of normal cellular metabolism, and that excessive free radical activity contributes to the pathogenesis of a variety of diseased states, and to the exacerbation of damage accompanying several aspects of physical trauma. Free radical biochemistry has become a rapidly expanding area of biological research, focusing strongly on the importance of several radical species in cellular injury, and on the possible value of free radical scavengers and antioxidants in the treatment of a wide range of degenerative disorders.

The present text has particularly emphasized the part played by nutrition in free radical defense, based on the recognition that several micronutrients form central components of the antioxidant defense system, and that, of all factors contributing to overall protection, they offer the most promising potential to be easily manipulated to the benefit of the individual. The approach taken in most chapters has been largely fundamental, with emphasis placed more on the mechanisms of free radical-driven cell damage and on the protective capacity of the micronutrients, than on their potential therapeutic value. Many questions remain unanswered at this level; these will need to be resolved before micronutrients can effectively be used for the protection of specific tissues against the unwanted activity of particular free radicals.

Chapters 1–6 develop the concept of natural and induced free radical damage within the cell, counterbalanced with varying degrees of effectiveness by a complex, multilevel, cellular-antioxidant defense system. The importance of controlled free-radical reactions in normal metabolism is stressed, as also are the serious consequences to cellular structures when control is lost. Factors aggravating peroxidative damage are highlighted and set against a background of changing antioxidant defense systems. In particular, the important point is made that within the cell, both the species of free radical produced, as well as the compounds used as antioxidants, differ markedly depending on subcellular location. More attention will need to be paid in the future to these distinctions, and to the cytoarchitectural distribution of antioxidants if antioxidant therapy is to be used effectively for the prevention and treatment of specific types of free radical damage and oxidative injury. This view is extended in the chapter on free radical pathology and the genome, which focuses on free radical production and DNA damage and defense within the nucleus.

Chapter 8 introduces a measure of speculation with respect to the involvement of free radicals in cancer and aging, and highlights the possible importance of damage to mitochondrial DNA in this context. Concluding with the paradox that because of its extended latency, cancer is really a disease of the young and middle aged, the author makes the plea that natural dietary antioxidants and supplements in certain circumstances be seen as important components in a preventative medical approach to modern health care. The last chapter raises the issue of free radical activity as a principal factor underlying the pathology of malnutrition, and is well supported by the wide clinical experience of the senior author. It is a fitting finale, as it focuses attention compellingly on the interaction of dietary micronutrients and free radical damage on one of man's longest enduring tragedies.

Free radicals have become fashionable in biological chemistry, and accordingly, do run the risk of being implicated too widely in cellular pathology. However, their existence has been proven experimentally, and their capacity to damage macromolecules cannot be denied. There can be little question that a large measure of degenerative change in the cell has its origins in free radical mechanisms. To the nutritional biochemist, the next challenge lies in establishing how antioxidant substances that occur naturally in the diet, or are administered as supplements, can best be used to diminish the extent of free radical-driven cellular damage, and the onset of degenerative change in the body. Apart from its broad potential value to community health, such research could find particular application with those individuals at special risk from free radical damage (e.g., patients undergoing extended X-irradiation or exposed to high levels of free-radical-generating xenobiotics).

Inevitably, progression into the field of antioxidant prophylaxis will lead to the development of synthetic antioxidants that may exceed in performance their natural counterparts. Increasingly, attention will focus on delivery systems that will optimize accumulation of the antioxidant into particular tissues, and will ensure the correct positioning of the compounds within the cytoarchitecture of the cell.

During the last decade, much progress has been made in defining the role of oxidants and antioxidants within the cell. The enormous potential of antioxidants in preventative medicine remains a largely un-tapped resource, but one that will ensure a high level of research activity in this area into the 21st century.

Index

A

Aging
   antioxidant defenses and, 43,
         191–191
   causes of, 43, 173–175
   free radicals and, 174, 186, 187
   lipoperoxidation and, 179
   lipopigments and, 43
Antioxidant defense system
   composition of, 107–126, 129,
         132–144, 134T, 135T,
         154, 160, 183–185,
         200–206
   malnutrition and, 201, 206
   trace elements and, 107–126,
         154, 183
   utilization of oxygen and, 183
   vitamins and, 129, 132–144, 154,
         184
Arthropathy
   collagen and, 38
Ascorbic acid
   antioxidant function of, 31, 108,
         134T, 138, 154, 161, 184,
         192, 203T
   prooxidant function of, 69, 139,
         208
   trolux and, 138

B

Bioflavinoids
   antioxidant function of, 135T, 143

Biogerontology
   preventative medicine in, 172
Biotin
   antioxidant function of, 135T,
         141

C

Calciferol
   antioxidant function of, 134T,
         136, 154, 203
Cancer
   causes of, 82, 159, 175–177
   free radicals and, 28, 159, 171,
         188–190
   mitochondria and, 189
   protection against, 159, 172,
         191–193
Carotenoids
   antioxidant function of, 29, 108,
         133, 134T, 138, 154, 184,
         203T
Catalase
   antioxidant function of, 27,
         110T, 116T, 154, 184,
         203T
Ceruloplasmin
   antioxidant function of, 31, 154,
         184
Choline
   antioxidant function of, 135T,
         142

deficiency
    carcinogenesis and, 82
    cell surface receptors and, 83
    ethanol toxicity and, 82
    fatty liver and, 81
Clinical conditions
    free radicals and, 78, 130, 182,
        188, 190, 191T
    lipid peroxidation and, 78
Cobalamin
    antioxidant function of, 134T,
        141
Coenzyme Q, *see* Ubiquinone
Copper
    antioxidant function of, 26, 109,
        110T, 154, 183, 203T
    deficiency of, 168, 189
    erythrocyte lifespan and, 112
    ethanol and, 113
    glutathione and, 111
    prooxidant function and, 151,
        182
Cytochrome oxidase
    superoxide production and, 26
Cytochrome P450
    superoxide production and, 56,
        59, 154, 187, 205

D

Deoxyribonucleic acid
    carcinogenesis and, 151, 161,
        175
    damage and, 149, 158, 174
    micronuclei and, 163
    mutagenesis and, 151, 159, 180
    protection and repair of,
        158–161
    transition metals and, 152, 155

E

Electron spin resonance, 2–11
Ethanol
    ethanol dehydrogenase and, 86
    ethanol free radical, 89
    hydroxyl radical from, 88
    iron and, 87, 94
    liver damage and, 84
    microsomal ethanol oxidizing
        system and, 87
    zinc and, 118

F

Fatty liver, 80–84
Fenton reaction, 61, 64, 155
Ferritin
    iron and, 94
    lipid peroxidation and, 92
Folacin
    antioxidant function of, 135T,
        140
Free radicals
    chain reactions of, 2
    definition of, 2, 26, 54, 150, 177,
        199
    detection of, 2, 5, 8
    half-lives of, 179
    inorganic radicals, 13, 18, 21,
        131
    metal chelates and, 37, 62, 64,
        69, 215
    organic radicals, 19, 56, 131
    production of, 4, 32, 54, 57, 107,
        130, 151, 154, 180, 200
    site specific reactions of, 65,
        152, 155
    transition metals and, 68, 151,
        155, 182, 215

G

Glutathione
   antioxidant function of, 31, 113,
      154, 184, 187, 205, 208
Glutathione peroxidase
   antioxidant function of, 28, 30,
      154, 157, 184, 203T, 207,
      209
   copper, influence of, 109
Glutathione-S-transferase
   antioxidant function of, 30, 205,
      210

H

Haber Weiss reaction, 35, 63, 155
Harcroft and Porta's concept, 78
Heme
   prooxidant function of, 59
Hydrogen peroxide
   prooxidant function of, 27, 32,
      36, 54, 132, 155
   sources of, 36
Hydroxyl radical, 17, 36, 54, 63,
      66, 87, 131, 155, 179, 202
Hyperoxia, 29
Hypoxia, 29

I

Inflammatory cells
   reperfusion injury and, 40
   respiratory burst and, 34, 38,
      200
Inositol
   antioxidant function of, 135T,
      142
Ionizing radiation
   DNA damage and, 153
   free radical production and,
      153, 164

lipid peroxidation and, 65
Iron
   autoxidation catalyzed by, 60
   desferrioxamine and, 68, 215
   ferritin as prooxidant and, 68,
      92, 94
   haptoglobin as antioxidant
      and, 68
   hemoglobin as prooxidant
      and, 56, 58, 68
   hemosiderin as prooxidant
      and, 69
   kwashiorkor and, 215
   lactoferrin as antioxidant
      and, 67
   myoglobin as prooxidant, 56
   nutritional overload of,
      91–94
   perferryl radical and, 67
   physiological forms of, 54
   prooxidant function of, 35, 55,
      60–69, 151, 155, 208
   transferrin as antioxidant
      and, 55, 67
   transferrin as prooxidant
      and, 94

K

Kwashiorkor
   free radicals and, 208–213
   iron stress in, 215
   prooxidant stress and, 213
   selenium and, 210
   sulfur amino acids and, 208
   symptoms of, 207

L

Lipid hydroperoxides, 131

Lipid peroxidation
  acetaldehyde and, 86
  carcinogenesis and, 82
  cell damage and, 91, 108,
      155, 179
  copper deficiency and, 111
  ethanol and, 85–90, 118
  iron and, 65, 91
  orotic acid and, 96
  pathological conditions
      and, 78, 182
  polyunsaturated fatty acids
      and, 79–84, 179
  radiation–induced, 65
  zinc deficiency and, 115
Lipoic acid
  antioxidant function of,
      135T, 142
Lipoxygenase, 59, 205

M

Malnutrition
  definition of, 202
  free radicals and, 204–208
  kwashiorkor and, 207
Manganese
  antioxidant function of, 27, 108,
      120, 122T, 155, 183, 203T
  complexes of, 120
  deficiency of, 121
  hydroxyl radical production
      and, 151
  lipoperoxidation and, 121
  mitochondrial membranes
      and, 121
  prooxidant function of, 151

Metallothionein
  antioxidant function of, 115,
      118, 154, 162
  zinc donation by, 118
Micronuclei, 163
Mitochondria
  aging and, 186
  cancer and, 189
  DNA damage in, 175
Mutagenesis, 158
Myeloperoxidase, 34
  hypochlorous acid and, 35, 55,
      138
Niacin
  antioxidant function of, 134T,
      140, 203T, 205
Nitrosamines
  ascorbic acid and, 192

O
Orotic acid, 94

P
Pantothenic acid
  antioxidant function of, 134T,
      141
Para-aminobenzoic acid
  antioxidant funcion of, 135T,
      143
Polyunsaturted fatty acids
  peroxidation of, 134T, 140

R
Retinol
  antioxidant funtion of, 133,
      134T, 154, 157, 203T
Reperfusion injury, 38, 40, 56

Riboflavin
   antioxidant function of, 134T,
      139, 203T

S
Selenium
   antioxidant function of, 27, 108,
      154, 157, 185, 203T
   kwashiorkor and, 210
Singlet oxyen, 11, 14, 26, 30, 136,
      138
Superoxide radical, 14, 32, 38, 55,
      58, 63, 111, 120, 131, 155,
      179, 202
Superoxide dismutase, 26, 56,
      109, 116, 154, 183, 203T
   ethanol and, 113, 123
   ozone and, 123

T
Thiamin
   antioxidant function of, 134T,
      139, 203T
Tocopherol
   antioxidant function of, 29, 92,
      108, 134T, 137, 154, 161,
      203T
Tranferrin
   iron release from, 94

U
Ubiquinone
   antioxidant function of, 135T,
      142
   superoxide production by, 33,
      39, 56, 135T, 142
UV-irradiation

   free radical production and,
      16, 65
   lipid peroxidation and, 65

V
Vitamin K
   antioxidant function of, 134T,
      138

X
Xanthine oxidase, 33
   reperfusion injury and, 39, 41,
      56

Z
Zinc
   antioxidant function of, 27, 108,
      114, 116T, 154, 161, 183,
      203T
   deficiency and, 114, 117
   ethanol and, 118
   interactions with other metals,
      115–117
   metallothionein and, 115, 118,
      162
   redox-cycling ions and, 115,
      161